£35-00

ANALYTICAL AND COMPUTATIONAL METHODS IN ENGINEERING ROCK MECHANICS

ANALYTICAL AND COMPUTATIONAL METHODS IN ENGINEERING ROCK MECHANICS

EDITED BY
E.T. BROWN

Imperial College of Science and Technology, London

London
ALLEN & UNWIN
Boston   Sydney   Wellington

Allen & Unwin (Publishers) Ltd,
40 Museum Street, London WC1A 1LU, UK

Allen & Unwin (Publishers) Ltd,
Park Lane, Hemel Hempstead, Herts HP2 4TE, UK

Allen & Unwin, Inc.,
8 Winchester Place, Winchester, Mass. 01890, USA

Allen & Unwin (Australia) Ltd,
8 Napier Street, North Sydney, NSW 2060, Australia

Allen & Unwin (New Zealand) Ltd,
Private Bag, Wellington, New Zealand

First published in 1987

British Library Cataloguing in Publication Data

Analytical and computational methods in engineering
rock mechanics.
1. Rock mechanics—Mathematics
2. Numerical calculations
I. Brown, E. T.
624.1'5132'072   TA706
ISBN 0-04-620020-7

Library of Congress Cataloging in Publication Data

Analytical and computational methods in engineering
rock mechanics
"Expanded version of the lectures given at the John
Bray Colloquium"—Pref.
Includes bibliographies and index.
1. Rock mechanics.   2. Brown, E. T.   I. Brown, E. T.
II. Bray, John.   III. John Bray Colloquium (1985:
Imperial College of Science and Technology)
TA706.A56   1986   624.1'5132   86-17477
ISBN 0-04-620020-7 (alk. paper)

Set in 10 on 12 point Times by Paston Press, Norwich
and printed in Great Britain by Mackays of Chatham

Preface

John Wade Bray was born in Bristol, England, on 26 March 1923. He studied civil engineering as an articled pupil obtaining a First Class Honours degree from the University of London in 1943. On graduation, he became a lecturer in the Department of Civil Engineering of the Northampton Polytechnic (now The City University). He was a research student at King's College London from 1947 to 1950 and was awarded his PhD for his research on temperature distributions in mass concrete. This work is reported in his first published paper (Ross & Bray 1949).* On completing his PhD, he returned to teaching, initially at Queen Mary College, then at the Northampton Polytechnic, and from 1953 at Imperial College where he has remained ever since.

As a lecturer in what was then the Mining Department of Imperial College, John Bray taught the basic engineering sciences to mining engineering undergraduates. At that time there was no specialist academic discipline of rock mechanics and no subject of that name in the mining curriculum. However, it was natural that he should approach the problems then of interest to his new colleagues by applying the principles and techniques of engineering mechanics. This effectively introduced rock mechanics into the teaching and research of the department. His personal research was not restricted to this new interest. The first paper that he published after joining the College (Bray 1957) describes a pioneering electrical analogue method of analysing rigid frameworks. It is interesting to note that his paper on temperature distributions in mass concrete (Ross & Bray 1949) and a subsequent paper on ventilation analysis (Bray & Plummer 1964) also present original methods of obtaining numerical solutions to important engineering problems. Although he subsequently used his formidable analytical powers to solve a wide range of problems of engineering rock mechanics interest, he retained throughout his career this early interest in developing efficient numerical methods for use in cases for which even he could not find analytical solutions.

John Bray's early research in rock mechanics was concerned with the stresses, displacements, and yield zones induced around underground excavations (Low & Bray 1962, Bray 1963–4, 1964, 1967c). However, he soon developed an interest in the influence of discontinuities on rock mass

* A list of John Bray's publications is given in chronological order at the end of this Preface.

behaviour and on the stability of excavations in rock. It was characteristic of him that he did not rush into print in these early days as others might have tended to do. However, in the mid-1960s, he wrote a series of seminal papers on the mechanics of discontinuous rock masses (Bray 1966, 1967 a & b); these papers retain an important place in the literature of the subject.

In 1966, the Inter-departmental Rock Mechanics Project was established at Imperial College under the leadership of Dr Evert Hoek. John Bray, who had been promoted to a Senior Lectureship in 1966, joined the Rock Mechanics Project in 1970. He became Director of the highly successful MSc Course in Engineering Rock Mechanics, and made significant contributions to the industrially funded research project on rock slope stability set up by Hoek. This work is reported in a number of papers (Hoek *et al*. 1973, Bray & Brown 1976, Goodman & Bray 1977), but most importantly in the book *Rock slope engineering* published in 1974 with second and third editions in 1977 and 1981. In 1979, Hoek and Bray were awarded the E. B. Burwell Award of the Geological Society of America for the contribution made by this book to 'the advancement of knowledge in engineering geology and the related field of rock mechanics'.

Although he developed original methods of analysis of a number of types of slope instability, John Bray continued to work on a range of fundamental problems in rock mechanics and on methods of analysis of underground excavations and foundations on rock. In the mid-1970s his interest in developing efficient numerical methods for rock mechanics application led him to an interest in boundary element methods which subsequently he shared with a number of colleagues and research students (Brady & Bray 1978 a & b, Austin *et al*. 1982, Crawford & Bray 1983). In 1980 Brady and Bray were given the Basic Research Award of the United States National Committee for Rock Mechanics for the original contribution made by their two papers on boundary element methods published in 1978. A continuing interest in plasticity theory was utilized in studies of rock mass properties, slope stability, foundations, and underground excavations (Brown & Bray 1982, Brown *et al*. 1983).

At some time or another, John Bray has studied almost every fundamental and applied problem of concern in engineering rock mechanics. Unfortunately, only a small proportion of his prolific output of original and often ingenious solutions has been published. Characteristically, having solved the problem at hand, he has always preferred to move on to the next challenge rather than writing up his work and, as he sees it, publicizing himself. Throughout his time in the Rock Mechanics Section (as it became), he would regularly appear at the college with neatly written out solutions to a problem that had been raised by a colleague, visitor, or student a few days before. The only outlets of much of his research have been his lectures to our MSc students and his conversations with our research students, a great number of whom have based their work on his ideas.

John Bray's work is characterized by a thorough knowledge of the basic engineering sciences, clarity of thought and presentation, and manipulative skills which few can hope to match. These characteristics are reflected in the organization and presentation of his lectures. He spends uncommonly large amounts of time on lecture preparation, preferring to develop his own derivations and methods of presenting results rather than relying on textbook presentations, many of which he finds unconvincing, if not wrong, and difficult for students to follow. Almost without exception, his many former students consider him to be the most outstanding of their university teachers.

John Bray is a reserved and modest man who has never sought the world's recognition or its material rewards. He does not enjoy travel and, throughout his career, declined regular invitations to speak at conferences or overseas universities and to undertake consultancies. It is highly probable that, had he wished it, he could have become head of the Rock Mechanics Section on more than one occasion. However, he has no entrepreneurial motivation whatever, preferring to spend his time on his research and teaching rather than on organizational, promotional, or administrative activities. In 1979, the University of London somewhat belatedly conferred on him the title of Reader in Rock Mechanics. On formally retiring in September 1984 (I say 'formally' because it can be guaranteed that he will never retire fully from the intellectual pursuits to which he has devoted his life), he was awarded the title of Emeritus Reader and became a Senior Research Fellow of Imperial College.

The esteem in which John Bray is held by his colleagues, students, and the rock mechanics community at large is such that we all wished to mark his formal retirement in some special way. It was readily agreed that the most appropriate way of doing so would be to hold a colloquium in his honour. Accordingly, about 50 of his past and present colleagues and research students from the Rock Mechanics Section gathered at Imperial College on 23 July 1985 for the John Bray Colloquium. State-of-the-art lectures on topics of current importance in engineering rock mechanics were presented by his distinguished former colleagues and/or students Barry Brady, Peter Cundall, Dick Goodman, and Evert Hoek, and by John Bray himself. It is important to note that the lecturers all agreed that the best way they knew of paying their personal tributes to John was to present a lecture representing the very best that they had to offer. The colloquium was a considerable academic and social success.

This book contains expanded versions of the lectures given at the John Bray Colloquium, together with an introductory chapter in which an attempt is made to put the subject of the book in perspective. The subject, analytical and computational methods in engineering rock mechanics, was chosen because it reflects accurately the nature of John Bray's work as well as being of central importance in rock mechanics. The book is not intended to be a textbook or to provide a comprehensive coverage of the subject. Rather, it focuses on a number of distinctive analytical and computational methods of special importance in engineering rock mechanics. It is anticipated that it will be used by teachers, advanced students, research workers, and advanced practitioners of the subject. All royalty income from the sales of the book will be used to provide a prize to be awarded annually to an outstanding student on the Imperial College MSc course in Engineering Rock Mechanics which John Bray directed with distinction from 1970 to 1984.

E. T. BROWN
London
February 1986

Publications of Dr J. W. Bray

1949   Ross, A. D. and J. W. Bray. The prediction of temperatures in mass concrete by numerical computation. *Mag. Concr. Res.* **1**, 9–20.

1957   Bray, J. W. An electrical analyser for rigid frameworks. *The Structural Engineer* **30**, 297–311.

1962   Low, I. A. B. and J. W. Bray. Strain analysis using moiré fringes. *The Engineer* **213**, 566–9.

1963–4   Bray, J. W. Discussion on 'Some problems of strata control and support in pillar workings' by A. Bryan and J. G. Bryan. *Trans. Instn Min. Engrs* **123**, 262–3.

1964   Bray, J. W. Discussion on 'A dynamic or energy approach to strata control theory and practice' by R. A. L. Black and A. M. Starfield. *Proc. 4th int. conf. strata control rock mechs*, Columbia University, New York, 463–7.

1964   Bray, J. W. and I. M. Plummer. Methods of ventilation analysis. *Min. Mag.* **110**, 224–37.

1966   Bray, J. W. Limiting equilibrium of fractured and jointed rocks. *Proc. 1st Congr. Int. Soc. Rock Mech.*, Lisbon, **1**, 531–5. Lisbon: Laboratório Nacional de Engenharia Civil.

1967a   Bray, J. W. A study of jointed and fractured rock. Part 1: Fracture patterns and their failure characteristics. *Rock Mech. Engng Geol.* **5**, 117–36.

1967b   Bray, J. W. A study of jointed and fractured rock. Part 2: Theory of limiting equilibrium. *Rock Mech. Engng Geol.*. **5**, 197–216.

1967c   Bray, J. W. The teaching of rock mechanics to mining engineers. *The Mining Engineer* **126**, 483–8.

1973   Hoek, E., J. W. Bray and J. M. Boyd. The stability of a rock slope containing a wedge resting on two intersecting discontinuities. *Q. J. Engng Geol.* **6**, 1–55.

1974   Hoek, E. and J. W. Bray. *Rock slope engineering*. London: Institution of Mining and Metallurgy. (2nd edn 1977; 3rd edn 1981.)

1976   Bray, J. W. and E. T. Brown. A short solution for the stability of a rock slope containing a tetrahedral wedge. *Int. J. Rock Mech. Min. Sci.* **13**, 227–9.

1977   Goodman, R. E. and J. W. Bray. Toppling of rock slopes. In *Rock engineering for foundations and slopes* **2**, 201–34. New York: American Society of Civil Engineers.

1978a   Brady, B. H. G. and J. W. Bray. The boundary element method for determining stresses and displacements around long openings in a triaxial stress field. *Int. J. Rock Mech. Min. Sci.* **15**, 21–8.

1978b   Brady, B. H. G. and J. W. Bray. The boundary element method for elastic analysis of tabular orebody extraction, assuming complete plane strain. *Int. J. Rock Mech. Min. Sci.* **15**, 29–37.

1978   Bray, J. W. Discussion on 'Energy changes due to mining' by J. B. Walsh. *Int. J. Rock Mech. Min. Sci.* **14**, 218.

1982   Brown, E. T. and J. W. Bray. Rock-support interaction calculations for pressure shafts and tunnels. In *Rock mechanics: caverns and pressure shafts*, W. Wittke (ed.), **2**, 555–65. Rotterdam: Balkema.

1982   Austin, M. W., J. W. Bray and A. M. Crawford. A comparison of two indirect boundary element formulations incorporating planes of weakness. *Int. J. Rock Mech. Min. Sci.* **19**, 339–44.

1983   Brown, E. T., J. W. Bray, B. Ladanyi and E. Hoek. Characteristic line calculations for rock tunnels. *J. Geotech. Engng, Am. Soc. Civ. Engrs* **109**, 15–39.

1983   Crawford, A. M. and J. W. Bray. Influence of the *in-situ* stress field and joint stiffness on rock wedge stability in underground openings. *Can. Geotech. J.* **20**, 276–87.

1986   Bray, J. W. Some applications of elastic theory. This volume, 32–94.

Acknowledgements

I wish to record my personal gratitude to the contributors Barry Brady, John Bray, Peter Cundall, Dick Goodman, Evert Hoek, and Gen hua Shi for their unhesitating and uncomplaining co-operation during the preparation of this book. I also wish to thank Moira Knox for her skilful work on the typescript, Laurie Wilson for his on the figures, and Robert Marsden for preparing the index.

The contributors wish to acknowledge the assistance of: colleagues at Imperial College, the University of Minnesota, and the CSIRO Division of Geomechanics, Australia (BHGB); the United States National Science Foundation for Grants CEE-8212674 and 8310729, Dr O. D. L. Strack of the University of Minnesota, Drs J. Jenkins and I. Ishibashi of Cornell University, and Bruce Trent, a PhD student at the University of Minnesota who made the three-dimensional plots of Section 4.4 on the computer system of the Los Alamos National Laboratory (PAC); and colleagues in Golder Associates, notably Ken Inouye and Trevor Fitzell, for help in developing the computer program listed in Appendix A, Chapter 3 (EH).

Finally, I wish to thank Roger Jones of Allen & Unwin for encouraging me to prepare this book for publication, and Geoff Palmer for the care and expertise that he brought to its presentation and production.

E.T.B.

Contents

List of tables

List of contributors

B. H. G. Brady Chief, Division of Geomechanics, CSIRO, Melbourne, Victoria, Australia

J. W. Bray Emeritus Reader in Rock Mechanics, Department of Mineral Resources Engineering, Imperial College of Science and Technology, London, UK

E. T. Brown Professor of Rock Mechanics and Head, Department of Mineral Resources Engineering, Imperial College of Science and Technology, London, UK

P. A. Cundall Associate Professor, Department of Civil and Mineral Engineering, University of Minnesota, Minneapolis, Minnesota, USA

R. E. Goodman Professor of Geological Engineering, Department of Civil Engineering, University of California, Berkeley, California, USA

E. Hoek Senior Principal, Golder Associates, Vancouver, British Columbia, Canada

Gen hua Shi Staff Scientist, Lawrence Berkeley Laboratory, University of California, Berkeley, California, USA

1 Introduction

E. T. BROWN

1.1 The nature of engineering rock mechanics problems

Rock mechanics has been defined as 'the theoretical and applied science of the mechanical behaviour of rock and rock masses; it is that branch of mechanics concerned with the response of rock and rock masses to the force fields of their physical environment' (Committee on Rock Mechanics 1966).

By making excavations in rock, or by constructing structures of or on rock, engineers change the force fields of the rock's physical environment. Thus, the subject of **engineering rock mechanics** is concerned with the response of rocks and rock masses to engineering activity. As in this book, the concern is usually with the stresses and displacements induced in the rock and with the stability of the structure. However, it must be recognized that the flow of fluids through rock masses is of central importance in a wide range of engineering rock mechanics applications. Heat transfer and its mechanical effects are also of concern in a limited number of applications; for some problems, notably the design of underground radioactive waste repositories, coupled thermal, hydraulic, and mechanical effects must be considered (Tsang *et al.* 1983).

The prediction of the engineering responses of rocks and rock masses is usually carried out within the framework of an overall design procedure. Figure 1.1 shows the components and logic of a widely accepted rock engineering design methodology. In this book our concern is with the methods used in the design analysis and retrospective analysis stages. It is important to recognize that the predictions made by the design analyses are functions of the geomechanical model of the rock mass developed from site characterization data. Even if the analytical or computational method chosen for the analysis is adequate, predictions of performance may be inaccurate because of deficiencies in the site characterization data or variability in the rock mass properties. For this reason, monitoring performance, retrospective analysis and updating the site characterization data and the geomechanical model are essential components of any effective engineering rock mechanics programme.

The difficulty of making reliable before the event or Class A predictions

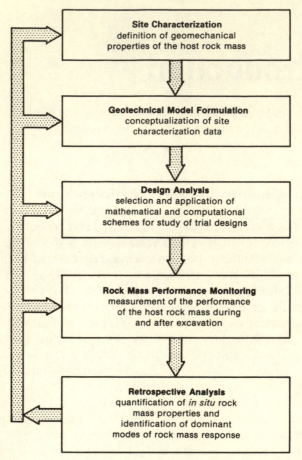

Figure 1.1   Components of a generalized engineering rock mechanics programme.

(Lambe 1973) of the engineering responses of rock masses derives largely from their discontinuous and variable nature. Rock masses contain bedding planes, joints, faults, and other structural features which render them discontinuous and often control their engineering behaviour. In addition, the response of the intact rock material itself may not be easy to model; it may have anisotropic and/or non-linear mechanical properties, it has low tensile strength, it is often susceptible to weathering, and its properties may change in the presence of water. Furthermore, the rock mass influencing the behaviour of a particular engineered structure may be a mechanically heterogeneous material consisting of a number of different rock types or of one rock type with variable properties. Because of the difficulty of assigning reliable values to a range of rock mass properties, parametric studies are widely used in design analyses (e.g. Hardy *et al*. 1979). For similar reasons, probabilistic approaches are being used increasingly in the description of rock masses and for engineering design in rock (e.g. Priest & Brown 1981).

1.2 Continuum and discontinuum models

Rock is distinguished from other engineering materials by the presence of discontinuities within it. Whether or not these discontinuities should be allowed for, either implicitly or explicitly, is the primary decision to be made in selecting an analytical or computational method for application to a particular rock mechanics problem. The development of methods of modelling discontinuities and their effects has been the distinctive feature of the adaptation to engineering rock mechanics of methods used in other branches of engineering mechanics.

Figure 1.2 shows a simplified representation of the influence exerted on the selection of a rock mass behaviour model by the relation between the discontinuity spacing and the size of the problem domain. A more detailed example is given by Brady in Figure 5.1 and described in Section 5.1. It may be that, on the scale of the problem, the rock mass is relatively free of discontinuities and may be treated as a continuum. Alternatively, the discontinuities may be so pervasive and closely spaced relative to the size of

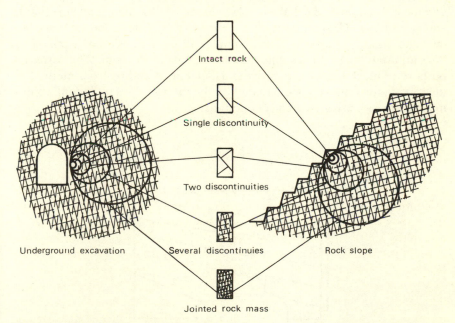

Intact rock

Single discontinuity

Two discontinuities

Underground excavation   Several discontinuies   Rock slope

Jointed rock mass

Figure 1.2 Simplified representation of the influence of scale on the type of rock mass behaviour model that should be used in designing underground excavations or rock slopes (after Hoek 1983).

the problem domain that the rock mass can be represented as a continuum with 'equivalent' rock mass properties. In either of these two cases, the classical **continuum theories** of **elasticity** and **plasticity** may be used. The elements of these theories are set out in Sections 1.3 and 1.4 below.

Sometimes, it may be necessary to allow for the presence of a single persistent discontinuity or a small number of discontinuities. This may be done, either analytically or computationally, by assuming rigid-body behaviour with slip on the discontinuities as in **limiting equilibrium methods**, or by calculating stress and displacement distributions using elastic theory and evaluating the potential for slip on the discontinuities under the imposed stresses.

An important class of problem remaining is that in which the problem domain is comprised of a finite number of discrete, interacting blocks. A number of methods have been developed for studying aspects of the resulting **discontinuum** behaviour (Trollope 1968, Cundall 1971, Stewart 1981, Goodman & Shi 1985, Lemos *et al*. 1985). The essential distinction between continuum and discontinuum behaviour is that, as illustrated in Figure 1.3, displacement fields need not be physically continuous in discontinua; individual blocks may be free to rotate or to translate with associated slip and/or separation at block interfaces. The most comprehensive discontinuum theory available, known as the **distinct element method**, is described by Cundall in Chapter 4 and used by Brady in a hybrid computational scheme in Chapter 5. **Block theory**, developed by Goodman and Shi (1985), and described in Chapter 6, uses the concepts of topology and set theory to evaluate important aspects of the complex three-dimensional geometry of discontinuous rock masses and excavations made in them. The essential purpose of block theory is to provide a convenient and rigorous method of identifying the critical blocks created by the intersections of discontinuities in a rock mass excavated along defined surfaces.

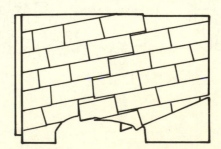

Figure 1.3   Discontinuum behaviour associated with an underground excavation in a blocky rock mass (after Voegele *et al*. 1978).

1.3 Elasticity

The primary continuum theory, the mathematical theory of elasticity (Love 1927), is used widely to calculate stresses, infinitesimal strains, and displacements induced in the rock following excavation or loading. Clearly, rocks and rock masses will not always behave elastically or as continua. Nevertheless, for a wide range of engineering problems, useful solutions may be obtained by treating the rock as a homogeneous, isotropic, linear elastic material (or HILE material as Bray has called it). If necessary, anisotropy (Gerrard 1977, Eissa 1980) and non-linear elasticity (Bray 1967a, Anderson & Jones 1985) may be allowed for.

In Chapter 2, Bray argues that, where applicable, elastic analysis can be used to evaluate a number of factors of importance in engineering rock mechanics. Examples given in respect of underground excavations are:

(a) the maximum and minimum stresses on the boundaries of openings;
(b) the boundary displacements induced by excavation;
(c) the extent of the zone of influence of an excavation;
(d) the extent of overstressed regions;
(e) the increase in stored strain energy, and the dynamic energy released, when an excavation is generated.

Most engineering rock mechanics applications of the theory of elasticity, including those presented in this book, are to static problems. Dynamic loading or unloading and wave propagation (as in studies of blasting mechanics) may also be analysed as problems in elasticity (see Jaeger & Cook 1979 or Brady & Brown 1985 for examples).

The conditions to be satisfied in obtaining a solution for the stress and displacement distributions in a particular case are:

(a) the boundary conditions of the problem;
(b) the differential equations of equilibrium;
(c) the constitutive equations for the material;
(d) the strain compatibility equations.

For the types of problem to be considered in this book, the boundary conditions are the *in situ* stresses in the rock mass and the tractions or displacements imposed on a surface of the rock mass by the excavation or structure. One of the major difficulties encountered in engineering rock mechanics is the determination of the stresses in the rock mass in the problem domain before excavation or construction. Unreasonable assumptions about the magnitudes and orientations of the *in situ* principal stresses, often made for analytical convenience, can lead to significant errors in the stresses calculated on the boundaries of underground excavations (Brady & St John 1982).

The differential equations of static equilibrium are

$$\frac{\partial \sigma_{xx}}{\partial x} + \frac{\partial \tau_{xy}}{\partial y} + \frac{\partial \tau_{zx}}{\partial z} + X = 0$$

$$\frac{\partial \tau_{xy}}{\partial x} + \frac{\partial \sigma_{yy}}{\partial y} + \frac{\partial \tau_{yz}}{\partial z} + Y = 0 \qquad (1.1)$$

$$\frac{\partial \tau_{zx}}{\partial x} + \frac{\partial \tau_{yz}}{\partial y} + \frac{\partial \sigma_{zz}}{\partial z} + Z = 0$$

where σ_{xx}, σ_{yy}, σ_{zz} and τ_{xy}, τ_{yz}, τ_{zx} are the direct and shear stress components in a three-dimensional x, y, z coordinate system, and X, Y, Z are the body forces per unit volume in the positive x, y, z directions.

The most general case of linear elastic constitutive behaviour is one in which any strain component is a linear function of all the stress components, i.e.

$$\{\varepsilon\} = [S]\{\sigma\}$$

or

$$\{\varepsilon\} = [D]^{-1}\{\sigma\} \qquad (1.2)$$

where $\{\varepsilon\}$ and $\{\sigma\}$ are column vectors of the six components of strain and stress, and the elements of the 6×6 matrices $[S]$ and $[D]$ are compliances and stiffness, respectively.

For an isotropic, linear elastic material, Equation 1.2 reduces to the well known statement of Hooke's law consisting of three expressions for direct strains of the form

$$\varepsilon_{xx} = \frac{1}{E} [\sigma_{xx} - \nu(\sigma_{yy} + \sigma_{zz})] \qquad (1.3)$$

and three for shear strains of the form

$$\gamma_{xy} = \tau_{xy}/G \qquad (1.4)$$

where E is Young's modulus, ν is Poisson's ratio and $G = E/2(1 + \nu)$ is the modulus of rigidity or shear modulus of the material.

It is a fundamental premise of the theory of elasticity that increments of stress and increments of strain are also uniquely related by equations of the form of Equation 1.2 with the elastic parameters being constants. Thus, elastic materials are path independent and exhibit conservation of energy or recovery of strain on unloading.

The strain compatibility equations arise from the requirement of physical continuity of the displacement field within a continuum. They consist of three equations of the form

$$\frac{\partial^2 \varepsilon_{xx}}{\partial y^2} + \frac{\partial^2 \varepsilon_{yy}}{\partial x^2} = \frac{\partial^2 \gamma_{xy}}{\partial x\, \partial y} \qquad (1.5)$$

and three of the form

$$2 \frac{\partial^2 \varepsilon_{xx}}{\partial y \, \partial z} = \frac{\partial}{\partial x} \left(-\frac{\partial \gamma_{yz}}{\partial x} + \frac{\partial \gamma_{zx}}{\partial y} + \frac{\partial \gamma_{xy}}{\partial z} \right) \tag{1.6}$$

In engineering rock mechanics, **plane strain** representations of problems are commonly used. If deformations are restricted to the x, y plane

$$\varepsilon_{zz} = \gamma_{yz} = \gamma_{zx} = 0$$

and the remaining components of strain (ε_{xx}, ε_{yy}, γ_{xy}) are not functions of z. In this case, and with zero body forces, Equations 1.1 and 1.3–1.6 may be combined to give

$$\left(\frac{\partial^2}{\partial x^2} + \frac{\partial^2}{\partial y^2} \right) (\sigma_{xx} + \sigma_{yy}) = 0 \tag{1.7}$$

Equation 1.7 demonstrates that two-dimensional stress distributions for isotropic elasticity are independent of the elastic properties of the medium, and that stress distributions for plane strain are the same as those for plane stress. This provides the justification for the use of plane stress photoelastic models in estimating stress distributions in plane strain problems such as those involving long underground excavations of constant cross section (Endersbee & Hofto 1963, Hoek 1967). As indicated by Equation 1.7, the sum of the plane normal stresses satisfies the Laplace equation. This provides the basis for the early use for an electrical analogue method in obtaining solutions for the elastic stresses and displacements around tabular mining excavations (Salamon *et al.* 1964).

To obtain analytical solutions for plane elasticity, the equilibrium equations and Equation 1.7 must be solved subject to the imposed boundary conditions. Airy (1862) introduced a stress function $U(x, y)$ such that

$$\sigma_{xx} = \frac{\partial^2 U}{\partial y^2}, \qquad \sigma_{yy} = \frac{\partial^2 U}{\partial x^2}, \qquad \tau_{xy} = -\frac{\partial^2 U}{\partial x \, \partial y} \tag{1.8}$$

These expressions satisfy the equilibrium equations for two dimensions and zero body forces. When substituted in Equation 1.7, they produce the biharmonic equation

$$\nabla^4 U = 0 \tag{1.9}$$

where

$$\nabla^2 = \frac{\partial^2}{\partial x^2} + \frac{\partial^2}{\partial y^2}$$

Several methods have been used to obtain solutions to particular problems in terms of Airy stress functions. For example, a polynomial of the form

$$U = \sum_{m=0}^{\infty} \sum_{n=0}^{\infty} C_{mn} x^m y^n$$

may be used in the case of continuous loads applied to the boundary. For axisymmetric problems formulated in cylindrical polar coordinates, a suitable stress function is of the form

$$U = A \ln r + Br^2 \ln r + Cr + D$$

where r is the radius coordinate.

For plane problems, complex variable theory provides the most elegant and powerful method of solution of the biharmonic equation. This approach is discussed in detail by Jaeger and Cook (1979). Further examples of its application to engineering rock mechanics problems are given by Salamon (1974) and by Bray in Chapter 2. The development of analytical solutions to problems in elasticity is aided considerably by the fact that the principle of superposition applies if all the differential equations and the boundary conditions are linear.

1.4 Plasticity

1.4.1 The incremental theory of plasticity

The incremental theory of plasticity (Hill 1950) is a branch of continuum mechanics that was developed in an attempt to model analytically the plastic deformation or flow of metals. Plastic deformation is permanent or irrecoverable. Perfectly plastic deformation occurs at constant volume under constant stress. If an increase in stress is required to produce further post-yield deformation, the material is said to be work- or strain-hardening.

Plastic or dissipative mechanisms of deformation may occur in rocks and rock masses on both the microscopic and macroscopic scales. Plastic mechanisms of deformation of rock material include cataclastic flow, intragranular slip and twinning (crystal plasticity), and diffusional flow (Paterson 1978). In rock masses, additional sources of plastic or irrecoverable deformation are slip on discontinuities and rotation and local crushing of rock blocks (Chappell 1974). Figure 1.4 shows that, in large-scale multiaxial compression tests on jointed biotite granite at the Kurobe IV arch dam site in Japan (Müller-Salzburg & Ge 1983), irrecoverable deformations developed at low levels of applied stress beyond an 'initial' yield limit. Subsequent 'principal' yield limits and peak or rupture stresses were also identified. The magnitudes of the irrecoverable components of deformation increased markedly above the principal yield limit. It would seem reasonable, therefore, to attempt to use plasticity theory to describe some aspects of the constitutive behaviour of rocks and rock masses.

Because plastic deformation is accompanied by permanent changes in atomic positions, plastic strains cannot be defined uniquely in terms of the current state of stress. Plastic strains depend on loading history, and so plasticity theory must use an incremental loading approach in which incre-

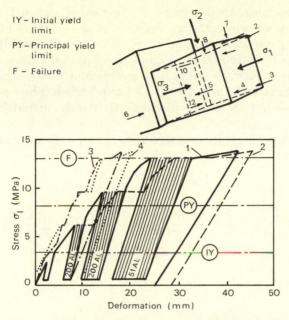

Figure 1.4 Major principal compressive stress–deformation curve for a multiaxial compression test, Kurobe IV arch dam site, Japan (after Müller-Salzburg & Ge 1983).

mental deformations are summed to obtain the total plastic deformation. In some engineering problems, the plastic strains are much larger than the elastic strains which may be neglected. This is not always the case for rock deformation (see, for example, Elliott & Brown 1985), and so an elasto-plastic analysis may be required. The total strain increment $\{\dot{\varepsilon}\}$ is the sum of the elastic and plastic strain increments:

$$\{\dot{\varepsilon}\} = \{\dot{\varepsilon}^e\} + \{\dot{\varepsilon}^p\} \tag{1.10}$$

A **plastic potential function**, $Q(\{\sigma\})$, is defined such that

$$\{\dot{\varepsilon}^p\} = \lambda \left\{\frac{\partial Q}{\partial \sigma}\right\} \tag{1.11}$$

where λ is a non-negative constant of proportionality which may vary throughout the loading history. Thus, from the incremental form of Equation 1.1 and Equations 1.10 and 1.11,

$$\{\dot{\varepsilon}\} = [D]^{-1}\{\dot{\sigma}\} + \lambda \left\{\frac{\partial Q}{\partial \sigma}\right\} \tag{1.12}$$

It is also necessary to be able to define the stress states at which yield will occur and plastic deformation will be initiated. For this purpose, a **yield function**, $F(\{\sigma\})$, is defined such that $F = 0$ at yield. If $Q = F$, the flow law is said to be **associated**. In this case, the vectors of $\{\sigma\}$ and $\{\dot{\varepsilon}^p\}$ are orthogonal; this is known as the **normality** condition. (This condition is illustrated in Figs 1.6 and 1.7). Bray (1982) notes that, for the von Mises yield criterion, $F(J_2) = 0$ where J_2 is the second invariant of the stress deviator tensor, the relation between stress and incremental plastic strain components are of the same form as the Saint-Venant relationships which are based on the Navier–Poisson law for perfectly viscous materials (Saint-Venant 1870).

For isotropic hardening and associated flow, elastoplastic stress and strain increments may be related by the equation

$$\{\dot{\varepsilon}\} = [D^{ep}]\{\dot{\varepsilon}\}$$

where

$$[D^{ep}] = [D] - \cfrac{[D]\left\{\cfrac{\partial Q}{\partial \sigma}\right\}\left\{\cfrac{\partial F}{\partial \sigma}\right\}^T [D]}{A + \left\{\cfrac{\partial F}{\partial \sigma}\right\}^T [D]\left\{\cfrac{\partial Q}{\partial \sigma}\right\}}$$

in which

$$A = -\frac{1}{\lambda}\frac{\partial F}{\partial K}\, dK$$

where K is a hardening parameter such that yielding occurs when

$$dF = \left\{\frac{\partial F}{\partial \sigma}\right\}^T \{\dot{\sigma}\} + \frac{\partial F}{\partial K}\, dK = 0$$

The concepts of associated plastic flow were developed for perfectly plastic and strain-hardening metals using yield functions such as those of Tresca and von Mises which are independent of the hydrostatic component of stress (Hill 1950). Although these concepts have been found to apply to some geological materials, it cannot be assumed that they will apply necessarily to pressure-sensitive materials such as rocks in which brittle fracture and dilatancy typically occur (Rudnicki & Rice 1975, Desai 1980).

Rocks and rock masses often display apparently strain-softening characteristics. The modelling of strain-softening behaviour using plasticity theory presents a number of difficulties. Plasticity is a continuum theory, but strain-softening in an isotropic continuum is impossible theoretically because it introduces instability. Strain-softening can exist only in a heterogeneous material (Bazant 1976). Heterogeneity in an initially homogeneous material that is deformed uniformly is produced by the localization of shear strain or fracture (Rudnicki & Rice 1975, Bazant 1976). Non-normality and non-uniqueness of solution may be associated with such behaviour. Although there are major difficulties involved, efforts to model shear strain and fracture localization in rock using plasticity theory (Vardoulakis 1983, 1984) seem to be well worth while.

In order to obtain realistic representations of the stresses at yield in rocks and rock masses, it has been necessary to develop yield functions which are more complex than the classical functions introduced for metals. Following Drucker and Prager (1952), these functions are often of the form $F(I_1, J_2)$ = 0 where I_1 is the first invariant of the stress tensor. It has been found experimentally that an assumption of associated flow overestimates the amount of dilation occurring in yielding rocks (see, for example, Michelis & Brown 1986). This observation has led to the development of a number of non-associated flow rules for rocks and rock-like materials (Maier & Hueckel 1979, Michelis & Brown 1986).

1.4.2   Limit theory*

Consider the elastic–plastic behaviour of the ground underlying a footing carrying an eccentric load P which is increased continuously from zero (Fig. 1.5). Under low loads, the ground is assumed to behave elastically with the vertical displacement and rotation of the footing base being relatively small. At a critical value of the load, the ground yields at A and a small plastic zone forms (Fig. 1.5a). Deformation is plastic within this zone but remains elastic outside it. As the load is increased, the plastic zone grows as in Figure 1.5b

* The development presented here is based on that used by J. W. Bray in an unpublished note (Bray 1977).

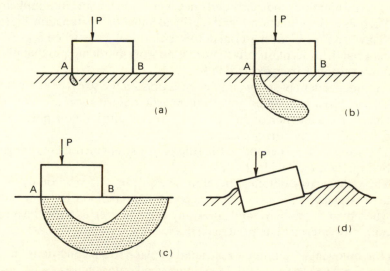

Figure 1.5   Collapse of a footing under increasing eccentric loading (after Bray 1977).

and, although displacements are greater than those resulting from purely elastic deformation, they are kept small by the restraining influence of the elastic material surrounding the plastic zone. Eventually, the plastic zone breaks through to the surface, as in Figure 1.5c; the system then becomes unstable with a large-scale rotation of the foundation base occurring. The value of the load at which this occurs is called the **collapse load**, P_c. Until this final large-scale rotation takes place, the geometry of the system varies little from that at zero load.

The determination of the complete elastic–plastic loading history using the incremental theory of plasticity and the basic equations set out in Section 1.4.1 can be an onerous task. However, it is possible to estimate the collapse load quite simply using **limit theory**. This theory permits the collapse load to be bracketed between two values called the upper and lower bounds (P_U, P_L) thus:

$$P_U > P_C > P_L$$

In most cases, upper and lower bounds can be established which are sufficiently close for the mean value to provide a reasonable estimate of the collapse load.

The following assumptions are made in the development and application of the theory.

(a) During plastic deformation, the geometry of the structure remains sensibly constant. This geometry is usually taken to be that in the unloaded state.
(b) Elastic deformations are negligible compared with plastic deformations. (This assumption is not absolutely necessary, but is made for simplicity.)
(c) There is a yield function F and a critical value of this function k^2, such that when $F < k^2$ the material deforms elastically, and when $F = k^2$ the material deforms plastically. There is no work-hardening so that plastic flow prevents F from exceeding k^2.
(d) F is independent of strain and depends only on the stress components.
(e) The material properties governing plastic deformation are isotropic. Hence, F is a function of the principal stresses, and is independent of the orientation of the principal axes.
(f) The flow rule is associated so that the plastic strain increments are given by Equation 1.11, with $Q = F$.
(g) The **yield surface** obtained by plotting $F(\sigma_1, \sigma_2, \sigma_3) = k^2$ on $\sigma_1, \sigma_2, \sigma_3$ axes is planar or convex. It is necessary to abandon the convention whereby $\sigma_1 > \sigma_2 > \sigma_3$, and allow any of these stresses to be the major, minor, or intermediate principal stress.

In the following, attention is restricted to plane strain conditions and the material is assumed to satisfy the Coulomb yield criterion, i.e.

$$|\sigma_1 - \sigma_3| - (\sigma_1 + \sigma_3) \sin \phi = 2c \cos \phi \qquad (1.13)$$

where c and ϕ are the cohesion and angle of internal friction, respectively. The yield function is

$$F = |\sigma_1 - \sigma_3| - (\sigma_1 + \sigma_3) \sin \phi \qquad (1.14)$$

and the critical value is

$$k^2 = 2c \cos \phi \qquad (1.15)$$

When $\sigma_1 > \sigma_3$, Equation 1.13 becomes

$$(\sigma_1 - \sigma_3) - (\sigma_1 + \sigma_3) \sin \phi = 2c \cos \phi \qquad (1.16)$$

and when $\sigma_3 > \sigma_1$, it becomes

$$(\sigma_3 - \sigma_1) - (\sigma_1 + \sigma_3) \sin \phi = 2c \cos \phi \qquad (1.17)$$

When plotted on σ_1, σ_3 axes, Equations 1.16 and 1.17 give the two continuous **yield lines** (AB, AC) shown in Figure 1.6.

Equations 1.11 and 1.14 give the principal plastic strain rates as

(a) when $\sigma_1 < \sigma_3$,

$$\dot{\varepsilon}_1 = \lambda(1 - \sin \phi)$$

$$\dot{\varepsilon}_3 = -\lambda(1 + \sin \phi)$$

or

$$\frac{\dot{\varepsilon}_1}{\dot{\varepsilon}_3} = -\left(\frac{1 - \sin \phi}{1 + \sin \phi}\right) \qquad (1.18)$$

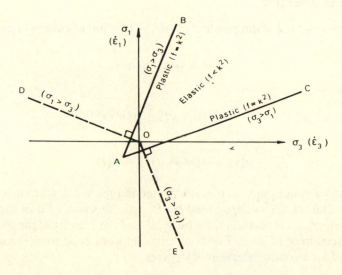

Figure 1.6 Coulomb yield with associated flow under plane strain conditions (after Bray 1977).

(b) when $\sigma_3 < \sigma_1$,

$$\dot{\varepsilon}_1 = -(1 + \sin \phi)$$

$$\dot{\varepsilon}_3 = (1 - \sin \phi)$$

or

$$\frac{\dot{\varepsilon}_1}{\dot{\varepsilon}_3} = -\left(\frac{1 + \sin \phi}{1 - \sin \phi}\right) \tag{1.19}$$

Plotting $\dot{\varepsilon}_1$ and $\dot{\varepsilon}_3$ on the σ_1 and σ_3 axes respectively, Equations 1.18 and 1.19 give the two broken lines (OD, OE) in Figure 1.6. It can be readily shown from Equations 1.16–1.19 that the pairs of lines AB, OD, and AC, OE are orthogonal. This is an illustration of the **normality** condition which is a consequence of the associated flow rule.

In Figure 1.6, a state of stress may be represented by either a point or a vector. Consider a state of plastic stress (σ_1^p, σ_3^p) represented by the vector $\boldsymbol{\sigma}^p$ in Figure 1.7. The corresponding rates of plastic strain are represented by a point on line OD in Figure 1.6 and by the components of a vector $\dot{\boldsymbol{\varepsilon}}^p$ in the direction of the outward normal to the yield line AB in Figure 1.7. Let another state of stress (σ_1^q, σ_3^q) be represented by the vector $\boldsymbol{\sigma}^q$. Then the vector difference $(\boldsymbol{\sigma}^p - \boldsymbol{\sigma}^q)$ will have

(a) a positive component in the direction of the outward normal to the yield line when $\boldsymbol{\sigma}^q$ represents an elastic state of stress (Fig. 1.7a),
(b) a zero component in the normal direction when $\boldsymbol{\sigma}^q$ represents a plastic state of stress (Fig. 1.7b).

It follows that the scalar product of $(\boldsymbol{\sigma}^p - \boldsymbol{\sigma}^q)$ and $\dot{\boldsymbol{\varepsilon}}^p$ cannot be negative, i.e.

$$(\boldsymbol{\sigma}^p - \boldsymbol{\sigma}^q)\dot{\boldsymbol{\varepsilon}}^p \geqslant 0$$

or

$$(\sigma_1^p - \sigma_1^q)\dot{\varepsilon}_1^p + (\sigma_3^p - \sigma_3^q)\dot{\varepsilon}_3^p \geqslant 0 \tag{1.20}$$

or

$$\sigma_1^p \dot{\varepsilon}_1^p + \sigma_3^p \dot{\varepsilon}_3^p \geqslant \sigma_1^q \dot{\varepsilon}_1^p + \sigma_3^q \dot{\varepsilon}_3^p \tag{1.21}$$

Consider a system such as that illustrated in Figure 1.5 which is loaded to collapse. Denote the collapse load by P^c and the stress field in the plastic zone by σ_1^c, σ_3^c. Let the strain rates be $\dot{\varepsilon}_1^c, \dot{\varepsilon}_3^c$ and the velocity of the load point in the direction of P^c be v^c. Equating external work done to internal energy dissipated for a volume element, dV, gives

$$P^c v^c = \int (\sigma_1^c \dot{\varepsilon}_1^c + \sigma_3^c \dot{\varepsilon}_3^c) \, dV \tag{1.22}$$

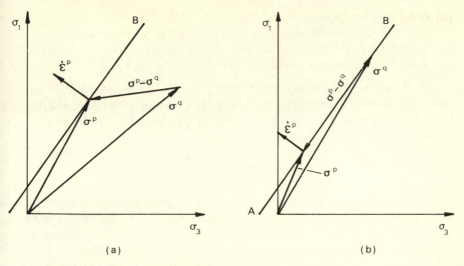

Figure 1.7 Relations between states of plastic stress and rates of plastic strain for yield under plane strain conditions (after Bray 1977).

The weight of the material could be included in this calculation but has been ignored for the sake of simplicity.

Consider now an imaginary stress system (σ_1^s, σ_3^s) called a **statically admissible stress field**, which is defined such that

(a) it satisfies the equations of equilibrium everywhere, both internally and on the boundaries, and at the load point it supports an external load P^s;
(b) it nowhere contravenes the yield criterion $F < k^2$.

Applying the equation of virtual work to this system and that at the collapse state gives

$$P^s v^c = \int (\sigma_1^s \dot{\varepsilon}_1^c + \sigma_3^s \dot{\varepsilon}_3^c)\, dV \qquad (1.23)$$

Subtracting this from Equation 1.22 gives

$$(P^c - P^s)v^c = \int [(\sigma_1^c - \sigma_1^s)\dot{\varepsilon}_1^c + (\sigma_3^c - \sigma_3^s)\dot{\varepsilon}_3^c]\, dV$$

By Equation 1.20, the integrand on the right-hand side is greater than or equal to zero. Thus,

$$P^c > P^s \qquad (1.24)$$

Hence P^s provides a **lower bound** to the collapse load P^c.

Consider now an imaginary system of velocities, called a **kinematically admissible velocity field** defined such that it

(a) provides a mechanism of collapse,
(b) excludes the development of gaps or overlaps in the material, and
(c) provides a velocity at the load point which has a positive component v^k in the direction of the load P^k.

Let σ_1^k, σ_3^k be the stresses which satisfy the yield criterion $F = k^2$ and which are related to the plastic strain rates $\dot{\varepsilon}_1^k$, $\dot{\varepsilon}_3^k$ through the associated flow rule. These stresses are not necessarily in equilibrium. The load P^k is calculated from the equation

$$P^k v^k = \int (\sigma_1^k \dot{\varepsilon}_1^k + \sigma_3^k \dot{\varepsilon}_3^k)\, dV \qquad (1.25)$$

Considering the collapse load P^c and stresses σ_1^c, σ_3^c in conjunction with the kinematically admissible velocity and strain rate field, the virtual work equation gives

$$P^c v^k = \int (\sigma_1^c \dot{\varepsilon}_1^k + \sigma_3^c \dot{\varepsilon}_3^k)\, dV \qquad (1.26)$$

Subtracting this from 1.25 gives

$$(P^k - P^c) v^k = \int [(\sigma_1^k - \sigma_1^c)\dot{\varepsilon}_1^k + (\sigma_3^k - \sigma_3^c)\dot{\varepsilon}_3^k]\, dV$$

By Equation 1.20 the integrand on the right-hand side is greater than or equal to zero. Thus,

$$P^k > P^c$$

Hence P^k provides an **upper bound** to the collapse load P^c.

This development establishes the **limit theorems** of the theory of plasticity which may be stated as follows:

(a) **Upper bound theorem.** If an estimate of the plastic collapse load of a body is made by equating the internal rate of dissipation of energy to the rate at which external forces do work in *any* postulated mechanism of deformation of the body, the estimate will be either high, or correct.
(b) **Lower bound theorem.** If *any* stress distribution throughout the structure can be found which is everywhere in equilibrium internally and balances certain external loads and at the same time does not violate the yield condition, those loads will be carried safely by the structure.

The limit theorems have been established assuming associated flow. Indeed, Davis (1967) has demonstrated that the theorems do not hold when flow is non-associated. However, there is some evidence from analyses carried out by Cox (1963) that many problems of practical interest are sufficiently unrestricted in their boundary deformation conditions for an analysis assuming associated flow to give correct or nearly correct answers for the collapse load, even though the velocity fields for associated and non-associated flow may differ markedly.

The limit theorems are applicable to materials which are perfectly plastic or which exhibit only a small degree of work-hardening. They should not be applied to cases in which collapse is associated with localized brittle fracture or with some other marked change in the internal structure of the material, or to cases of mechanical instability involving major changes in geometry such as buckling or toppling. Nevertheless, Davis (1980) has shown that applications of the limit theorems of plasticity can provide some insight into the influence of strain-softening and of slip on discontinuities on the load-bearing capacities of fissured rocks and clays.

Although they were developed for ductile metals, the limit theorems have been applied to a wide range of soil and rock mechanics problems. Heyman (1972) has demonstrated that the approaches used by Galileo (1638) in his theory of the bending of beams and by Coulomb (1776) in his essay on 'some statical problems' can be brought within the framework of the limit theorems of the theory of plasticity. Coulomb considered initially the problem of the collapse load of a square stone column loaded in uniaxial compression (Fig. 1.8). Heyman notes that Coulomb's treatment of this problem is an upper bound approach; his postulate of a plane of failure CM, whose location is unknown at the start of the analysis is equivalent to the specification of a velocity field.

Limit theory may be used to determine the critical heights of slopes in uniform materials which obey a Coulomb yield criterion. It is found that a log spiral slip surface gives the best available upper bound solution (Drucker & Prager 1952). Bray (1977) presents an original and instructive approach to this problem. Assume that failure occurs by slip on a circular cylindrical

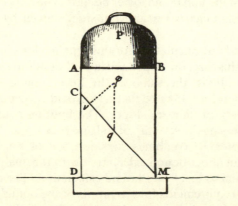

Figure 1.8 Square stone column loaded in uniaxial compression (after Coulomb 1776).

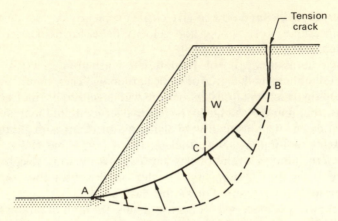

Figure 1.9   Circular slip surface in a drained, uniform slope (after Bray 1977).

surface as illustrated in Figure 1.9, and that there is zero excess pore pressure
throughout. An accurate value of the factor of safety against shear failure
could be determined if the stress distribution on the slip surface were known.
However, all that is known is that at limiting equilibrium the normal stresses
and the associated shear stresses on this surface must (a) balance the weight
of the sliding wedge, and (b) satisfy the yield criterion.

Bray (1977) found that there are two limiting normal stress distributions
for which the factor of safety against shear failure has its maximum and
minimum values or upper and lower bounds. In the first case, the normal
stress is concentrated as two line loads at the extreme ends of the slip surface,
A and B; in the second case it is concentrated as a single normal reaction at
C which lies on the line of action of W. Since the positions of the tension
crack and of the centre of curvature of the slip surface are not known, it is
necessary to repeat the calculations for all possible geometries to determine
minimum values of the upper and lower bounds. This procedure was used in
producing the design charts for uniform slopes given by Hoek and Bray
(1981).

Bray also carried out calculations of this type for log-spiral slip surfaces.
He found that, in this case, the upper and lower bounds to the factor of safety
are identical and almost the same as the lower bounds for circular slip
surfaces which, therefore, may be taken to provide acceptable estimates of
the factors of safety of uniform slopes. The limiting equilibrium method
widely used in assessing the stability of soil and rock slopes (as, for example,
by Hoek in Chapter 3), combines consideration of a proposed collapse
mechanism as in an upper bound calculation with the analysis of conditions
of static equilibrium as in a lower bound calculation. The method does not
satisfy the formal requirements for the proofs of the bound theorems, but it
has been found to give solutions which agree well with observations of slope
failures (Hoek & Bray 1981).

1.5 Analytical methods

In engineering mechanics, analysis consists of resolving a problem into its simple elements, representing the problem by tractable equations, and solving those equations (Gibson 1974). Analytical methods are generally regarded as being those which produce closed form or pseudo closed form solutions. In order to be tractable, they may require that simpler representations be made of problems than those used in computational schemes based on similar behavioural models. Gibson observed: 'The choice of what simplifications to make will be guided by our prior expectation of the likely effect of each on the predictions. It may be good tactics and sometimes the only feasible course open, to oversimplify the model and then study seriatum the influence of the discarded factors.' Making these simplifications requires a sound knowledge of engineering mechanics, imagination, and judgement, qualities which are amply demonstrated in Bray's work.

In his discussion of the analytical method in soil mechanics, Gibson (1974) concluded: 'The analytical method draws attention to broad trends and helps to distinguish between which factors are of primary significance and which are of secondary importance. It is neither a dispensable supplement to engineering intuition, nor merely a procedure for quantifying results. It is able to "speak for itself" and on occasion does so in characteristic ways.' As well as being used as a tool in the design process outlined in Figure 1.1, the analytical method also plays the rôle in engineering rock mechanics which Gibson identified for it in soil mechanics. Bray has used analysis to great effect in this sense in both teaching and research.

It is only rarely that complete analytical solutions can be found for the problem geometries, boundary conditions, and constitutive laws applying in practical engineering rock mechanics problems. However, it is often possible to obtain analytical solutions to problems which approximate to those of practical interest and from them to develop valuable insight into the problem of concern. For example, Brady and Brown (1985) show how the general plane strain solution for the distribution of elastic stresses around an elliptical excavation, developed by Bray in Section 2.3, may be used to estimate the zones of influence and the critical boundary stresses for excavations having a range of shapes. Even where analytical solutions to pertinent 'real world' problems cannot be found, analytical solutions to **singular problems** in which a disturbance is applied at a point in a homogeneous material can be of great value in developing numerical solutions.

The following are among the general categories of rock mechanics problems which have been solved analytically and which are of interest in the context of this book:

(a) Distributions of elastic stress and displacement induced around excavations of simple geometry. As noted in Section 1.3, these solutions are usually obtained using Airy stress functions or complex variable theory.

Examples of such solutions for isotropic and anisotropic materials are developed by Bray in Chapter 2.

(b) Distributions of elastic stress and displacement induced in isotropic and anisotropic materials by foundation loads (Poulos & Davis 1974).

(c) Determination of the extent of, and distribution of stress and displacement within, plastic zones formed around circular excavations subject to axisymmetric boundary conditions. Brown *et al.* (1983) review solutions obtained for a variety of material behaviour models.

(d) Limiting equilibrium analyses of the stability of individual blocks of rock isolated in slopes or on the boundaries of underground excavations (Hoek & Brown 1980, Hoek & Bray 1981).

(e) Determination of the upper and lower bounds to the collapse loads for foundations (Davis 1967, 1980), slopes (Drucker & Prager 1952, Hoek & Bray 1981), and underground excavations (Davis *et al.* 1980, Mühlhaus 1985) in $\phi = 0$ and, more importantly, c–ϕ materials.

1.6   Computational methods

1.6.1   Classification of methods

As noted above, it is only rarely that analytical solutions can be found to rock mechanics problems of practical concern. This may be because the boundary conditions cannot be described by simple mathematical functions, the governing partial differential equations are non-linear, the problem domain is inhomogeneous, or the constitutive relations for the rock mass are non-linear or otherwise insufficiently simple mathematically. In these cases, approximate solutions may be found using computer-based numerical methods.

In general, numerical methods of solving boundary value problems can be divided into two classes, those that require approximations to be made throughout the problem domain and those that require approximations to be made only on a boundary. The first group are known as **differential methods** and the second as **integral methods**. The boundary referred to in the case of integral methods may enclose the problem domain, in which case the problem is described as an **interior problem**. Of greater interest in engineering rock mechanics is the **exterior problem**, in which the problem domain is an infinite or semi-infinite region outside the boundary of concern. In a general sense, discontinuum methods fall within the broad class of differential methods. However, in the present context, they will be considered separately from differential continuum methods. Figure 1.10 illustrates the essential differences between differential and integral methods of modelling an exterior problem in a continuum.

A powerful group of methods of recent origin are those which combine differential and integral or continuum and discontinuum methods in solving a particular problem. These are referred to as **linked** or **hybrid methods**.

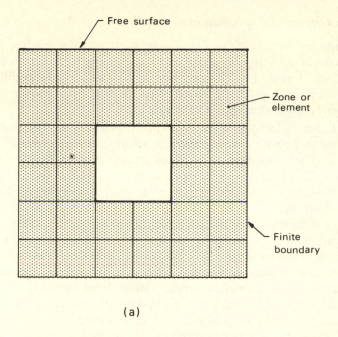

(a)

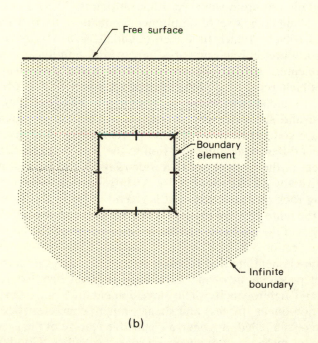

(b)

Figure 1.10 (a) Differential and (b) integral methods of modelling an exterior problem in a continuum (after Hardy *et al.* 1979).

1.6.2 Differential continuum methods

Differential continuum methods may be used to solve boundary value problems in elasticity and plasticity formulated using the concepts outlined in Sections 1.3 and 1.4. The methods in this group, the **finite difference** and the **finite element** methods, require that physical or mathematical approximations be made throughout a bounded region.

In finite difference methods, approximate numerical solutions are obtained to the governing equations at an array of points within the problem domain. Thus, this method provides an approximate solution to an exact problem. Brown *et al.* (1983) give a simple example of the application of a finite difference technique to the solution of an axisymmetric problem in elastoplasticity.

Finite difference methods are used only rarely in the solution of steady-state problems or problems in statics. However, they are used more widely in solving transient or dynamic problems, including those of engineering rock mechanics interest (Cundall 1976). As in the finite element method, problems may be formulated either implicitly or explicitly. In an **implicit** formulation, a set of equations is derived and solved to determine a new state at some given time. In an **explicit** formulation, the state of the system is, in effect, frozen at each time step and a new condition at each calculation point determined directly from values at adjacent points. The advantage of this procedure is that no new set of simultaneous equations need be established, stored, and solved at each time step (Cundall 1976, Hardy *et al.* 1979). Explicit procedures have particular advantages for non-linear problems.

In finite element methods, the problem domain is subdivided into discrete elements which provide a physical approximation to the continuity of displacements and stresses within the continuum. The governing equations are written and solved exactly for points, or nodes, at which adjacent elements are connected. Thus, the finite element method gives an exact solution to a differential approximation to the problem. Full details of the development of the method and of a wide range of its engineering applications are given by Zienkiewicz (1977). An introduction to the method in an engineering rock mechanics context is given by Brady and Brown (1985). Although the finite element method is not considered in detail in this book, the essentials of the mathematical formulation of the method are presented by Brady in Section 5.8.

The finite element method is better suited than integral methods to the solution of problems involving non-linear material behaviour, including plasticity and heterogeneity. The development of joint elements which allow the non-linear stiffness and shear strength characteristics of discontinuities to be modelled, has been a distinctive feature of the application of finite element methods in engineering rock mechanics. Goodman and St John (1977) set out the essential features of the adaptation of the method to the numerical modelling of the mechanical response of discontinuous rocks.

In addition to the fact that discretization errors can occur throughout the problem domain, the major limitation of the finite element method for application to the exterior problems commonly met in engineering rock mechanics lies in the need to define arbitrarily the outer boundary of the problem domain (see Fig. 1.10a). This can introduce inaccuracies into the solution because the far-field stress conditions may not be satisfied completely. If the problem domain is extended sufficiently to enable this difficulty to be alleviated, a cost penalty will be incurred through increased data preparation and computing times. The difficulty can be overcome by the use of infinite elements (Beer & Meek 1981) or by linking finite element and boundary element schemes. The latter approach, which is discussed by Brady in Section 5.8, combines the power and economy of integral methods in representing linear far-field behaviour with the versatility of the finite element method in modelling non-linear near-field response.

Numerical solutions to plane problems in plasticity may be found using a differential method known as the method of **characteristic lines**. This method assumes purely plastic behaviour and permits the use of a flow law which need not be associated. As in applications of limit theory, it is usually necessary to have some concept of the location and nature of the plastic zones in a specific problem before a solution can be attempted.

If, for plane problems, the relation between the stress components at yield according to Coulomb's criterion is substituted into the differential equations of equilibrium, a pair of quasi-linear hyperbolic equations results. Associated with this set of hyperbolic equations are two families of lines known as characteristics and usually referred to as α and β lines. The α and β lines intersecting at any point in a plastic zone are each inclined at an angle $\mu = \pi/4 - \phi/2$ to the major principal stress direction (Fig. 1.11). The pair of hyperbolic equations arising from the equilibrium equation and the yield function may be replaced by a second pair of equations, one of which

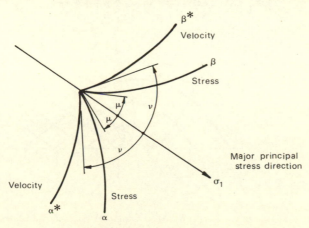

Figure 1.11   Definition of stress and velocity characteristics.

involves differentiation along α lines and the other differentiation along β lines. This representation is particularly useful in developing numerical solutions. Booker and Davis (1977) discuss a widely used iterative method of numerical integration of these stress equations.

In the general case, it is not possible to solve a problem in plasticity by considering only equilibrium; the velocity field must be determined as well. It is necessary, therefore, to check that the computed stress field is compatible with some reasonable form of movement and that the internal rate of doing plastic work is everywhere positive. It is found that families of velocity characteristics α^* and β^* may be defined inclined at angles ν to the major principal stress direction (Fig. 1.11). If the material obeys an associated flow rule, $\nu = \mu$ and the velocity and stress characteristics coincide.

It is interesting to note that in some of his earliest papers on engineering rock mechanics, Bray (1966, 1967b) made ingenious use of the characteristic line method. He compared two similar systems:

(a) a mass of material having a cohesion c and an angle of friction ϕ' loaded to a state of plastic equilibrium, and
(b) geometrically similar mass of fractured rock having the same cohesion, c, but a different angle of friction, ϕ, and loaded by the same system of external forces as (a).

He used the characteristic lines for (a) to determine the form of the fracture pattern which makes system (b) critical under these conditions and leads to the same distribution of stress. Figure 1.12 shows a solution obtained by

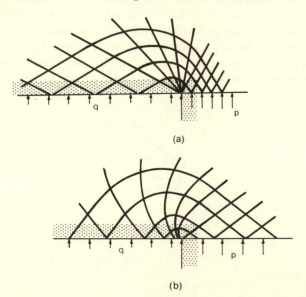

(a)

(b)

Figure 1.12 Fracture zone at the end of a stope: (a) characteristic lines for $\phi' = 30°$; (b) equivalent fracture system for rock with $\phi = 45°$. (After Bray 1967b.)

Bray for a stope in a horizontal tabular orebody. It is assumed that there is a uniform vertical abutment stress p and a uniform support pressure q acting on the hanging wall. The weight of the material in the fractured zone is ignored and the material is assumed to be cohesionless. Figure 1.12a shows the characteristic lines for $\phi' = 30°$; the corresponding limiting value of p/q is 18.4. Figure 1.12b shows the equivalent fracture pattern for rock for which $\phi = 45°$; the limiting value of p/q is now 134.

1.6.3 Integral methods

As noted in Section 1.6.1, integral methods make approximations only on the external boundary of a solid body in an interior problem or on an internal boundary in an exterior problem. As illustrated in Figure 1.10b, only the boundary is divided into elements. Numerical solutions use analytical solutions for simple singular problems in such a way as to satisfy approximately for each element the boundary conditions in terms of imposed tractions and displacements. Since the problem boundary only is defined and discretized, integral methods effectively provide a unit reduction in the dimensional order of a given problem. This reduces the size of the system of equations to be solved and offers significant advantages in computational efficiency over differential methods. This makes integral methods especially useful for solving the three-dimensional problems in elastostatics which are sometimes of concern in engineering rock mechanics (Hocking *et al.* 1976, Watson & Cowling 1985). Integral methods model far-field boundary conditions correctly and ensure continuous variation of stress and displacement throughout the material. They are best suited to applications involving homogeneous materials and linear material behaviour.

Integral methods may use either direct or indirect formulations. In indirect formulations, the physical problem is replaced by an equivalent problem which has the same solution as the original problem. This approach makes use of 'fictitious forces' and 'fictitious displacements' on an excavation boundary, for example. Two indirect methods used in engineering rock mechanics are known as the **boundary element** method (see Hoek & Brown 1980 for a description and listing of a two-dimensional boundary element program developed by J. W. Bray) and the **displacement discontinuity** method (Salamon 1974, Crouch & Starfield 1983). The direct formulation as used by Watson (1979) makes use of Betti's reciprocal theorem to eliminate 'fictitious' forces and displacements; this is known as the **boundary integral equation** method. It should be noted that these integral methods are often referred to collectively as boundary element methods.

Details of these various methods and examples of their application to rock mechanics problems are given by Crouch and Starfield (1983). An account of the indirect boundary element formulation is given by Brady and Brown (1985). In Chapter 5, Brady presents the direct formulation for an elastic continuum and shows how it may be modified for application to non-homogeneous media and problems involving slip on dominant discontinuities. Brady also discusses the linkage of boundary element methods with finite element and distinct element schemes.

1.6.4   Discontinuum methods

Special methods of solution are required for those engineering rock mechanics problems involving the interaction of finite numbers of discrete blocks of rock and in which the ratio of block size to the size of the problem domain is such that equivalent continuum behaviour may not be assumed. Partial solutions to some problems in this class may be obtained by applying the equations of statics (Bray 1966; Trollope 1968; Goodman & Bray 1977). However, these approaches necessarily make restrictive assumptions about block geometries and the deformabilities of blocks and discontinuities and are unable to determine the magnitudes of block displacements. More complete solutions to less idealized problems require that special computational schemes for discontinua be developed.

Burman (1971) developed a finite element model for the simulation of the behaviour of discontinua composed of units of constant shape and size. This model allows for non-linear shear and normal deformation at unit contacts and represents deformations in terms of rigid-body displacements of adjacent unit centroids. Only small displacements are allowed for, and the method does not model several distinctive features of discontinuum response.

The most powerful and versatile method available for simulating discontinuum behaviour is the **distinct element** method originally developed by Cundall (1971) as a means of modelling the progressive failure of rock slopes. This method uses a dynamic relaxation technique to solve Newton's laws of motion to determine the forces between, and the displacements of, units during the progressive, large-scale deformation of discontinua. Cundall gives an account of the method in Chapter 4, and in Chapter 5 Brady shows how it may be linked to a boundary element scheme for application to underground excavation design. Accordingly, the basis and use of the method will not be discussed here. The original distinct element method assumed the units to be rigid. Figure 1.3 was produced using a version of the method incorporating rigid elements and developed for use in modelling the behaviour of the rock around underground excavations. The method has now been extended to include element deformability (Lemos *et al*. 1985).

In an attempt to reduce the computing time required for convergence in some problems, Stewart (1981) devised an alternative static relaxation scheme for modelling the discontinuous behaviour of rock around underground excavations. The static relaxation method seeks a direct force-displacement path to equilibrium, independent of the time-stepping integration procedure inherent in dynamic relaxation. In other respects, many of the assumptions and features of the two relaxation methods are similar. However, the static relaxation method has not yet been developed to the stage at which it can match the power and versatility of current distinct element codes.

1.6.5 Code verification

It is necessary that computer codes based on any of the methods outlined in the preceding sections be verified before being used in practice or in research. Code verification involves demonstrating, to an acceptable level of tolerance, correspondence between the code's solution to a problem and an independent solution to that problem. The following discussion of code verification follows that of Brady and St John (1982).

The level of effort required for code verification increases with the complexity of the constitutive behaviour incorporated in the computational scheme. Suitable tests must be defined for all segments of the code and for all circumstances likely to arise during its application. Unfortunately, it is frequently only the writer of a code who is sufficiently familiar with the subtleties of the code's construction to be able to design and execute an adequate programme of code verification tests.

The simplest codes to verify are those based on linear elasticity. Analytical solutions to a wide variety of problems provide suitable independent checks. A code should be accepted as being satisfactory only if it gives the required correspondence with an established analytical solution with a sensible number of elements. For example, a boundary element code which requires 100 elements on the boundary of a circular excavation to produce the desired level of agreement with the analytical solution could not be regarded as being satisfactory. A further problem in verifying linear codes may arise from the symmetry of the problems for which closed-form analytical solutions exist. In these cases, correspondence between the numerical and analytical solutions is a necessary but not sufficient condition for code verification. It is possible that symmetric problems can be processed in a code with self-cancelling internal errors. In such cases, the code must be checked against another code having a different analytical basis.

Verification of codes incorporating discontinuum or other forms of non-linear behaviour presents acute difficulty. The classical analytical solutions for problems in plasticity provide only interim steps in the verification of codes based on elastoplasticity. In the absence of reliable analytical solutions for the more complex constitutive behaviour of rock masses, the only logical approach is to conduct controlled physical model tests. Verification of the satisfactory performance of a code requires correspondence, to an agreed tolerance, between the observed behaviour of the test specimen and that given by the code for the boundary and loading conditions of the test.

Clearly, in non-linear problems, loading and computational paths may influence the results obtained; these features add to the challenge of validating non-linear codes. Field tests make little contribution to code verification because material properties, boundary conditions, and loading conditions cannot be defined, controlled, and reproduced sufficiently closely.

References

Airy, G. B. 1862. On the strains in the interior of beams. *Rep. 32nd Mtg Br. Assoc. Adv. Sci.*, Cambridge, 82–6.

Anderson, M. A. and F. O. Jones 1985. A comparison of hydrostatic-stress and uniaxial-strain pore-volume compressibilities using nonlinear elastic theory. In *Research and engineering applications in rock masses*, Proc. 26th US symp. rock mech., Rapid City, E. Ashworth (ed.), **1**, 403–10. Rotterdam: Balkema.

Bazant, Z. P. 1976. Instability, ductility and size effect in strain softening concrete. *J. Engng Mech. Div., Am. Soc. Civ. Engrs* **102**, 331–44.

Beer, G. and J. L. Meek 1981. Infinite domain elements. *Int. J. Numer. Meth. Engng* **17**, 43–52.

Booker, J. R. and E. H. Davis 1977. Stability analysis by plasticity theory. In *Numerical methods in geotechnical engineering*, C. S. Desai and J. T. Christian (eds), 719–48. New York: McGraw-Hill.

Brady, B. H. G. and E. T. Brown 1985. *Rock mechanics for underground mining*. London: Allen & Unwin.

Brady, B. H. G. and C. M. St John 1982. The role and credibility of computational methods in engineering rock mechanics. In *Issues in rock mechanics*, Proc. 23rd US symp. rock mech., Berkeley, R. E. Goodman and F. E. Heuzé (eds), 571–86. New York: Society of Mining Engineers, American Institute of Mining, Metallurgical and Petroleum Engineers.

Bray, J. W. 1966. Limiting equilibrium of fractured and jointed rocks. *Proc. 1st congr., Int. Soc. Rock Mech.*, Lisbon, **1**, 531–5. Lisbon: Laboratório Nacional de Engenharia Civil.

Bray, J. W. 1967a. The teaching of rock mechanics to mining engineers. *The Mining Engineer* **126**, 483–8.

Bray, J. W. 1967b. A study of jointed and fractured rock. Part 2: Theory of limiting equilibrium. *Rock Mech. Engng Geol.* **5**, 197–216.

Bray, J. W. 1977. *Limit theory – collapse loads, upper and lower bounds*. Unpublished note. Imperial College of Science and Technology, London.

Bray, J. W. 1982. *Roots of the associated flow rule*. Unpublished note. Imperial College of Science and Technology, London.

Brown, E. T., J. W. Bray, B. Ladanyi and E. Hoek 1983. Characteristic line calculations for rock tunnels. *J. Geotech. Engng, Am. Soc. Civ. Engrs* **109**, 15–39.

Burman, B. C. 1971. *A numerical approach to the mechanics of discontinua*. PhD thesis, James Cook University of North Queensland.

Chappell, B. A. 1974. Load distribution and deformational response in discontinua. *Géotechnique* **24**, 641–54.

Committee on Rock Mechanics 1966. *Rock Mechanics research: a survey of United States rock mechanics research to 1965, with a partial survey of Canadian universities*. Washington, DC: National Academy of Sciences.

Coulomb, C. A. 1776. Essai sur une application des règles de maximis et minimus à quelques problèmes de statique, relatifs à l'architecture. *Mémoires de Mathématique et de Physique, l'Académie Royale des Sciences* **7**, 343–82.

Cox, A. D. 1963. *The use of non-associated flow rules in soil plasticity.* Report (B) 2/63, Royal Armament Research and Development Establishment, Chertsey.

Crouch, S. L. and A. M. Starfield 1983. *Boundary element methods in solid mechanics.* London: Allen & Unwin.

Cundall, P. A. 1971. A computer model for simulating progressive large scale movements in blocky rock systems. In *Rock fracture*, Proc. int. symp. rock fracture, Nancy, Paper 2–8.

Cundall, P. 1976. Explicit finite-difference methods in geomechanics. In *Numerical methods in geomechanics*, C. S. Desai (ed.), **1**, 132–50. New York: American Society of Civil Engineers.

Davis, E. H. 1967. A discussion of theories of plasticity and limit analysis in relation to the failure of soil masses. *Proc. 5th Aust.–NZ conf. soil mech. foundn engng*, Auckland, 175–82.

Davis, E. H. 1980. Some plasticity solutions relevant to the bearing capacity of rock and fissured clay. *Proc. 3rd Aust.–NZ Conf. Geomech.*, Wellington, **3**, 3.27–3.36.

Davis, E. H., M. J. Gunn, R. J. Mair and H. N. Seneviratne 1980. The stability of shallow tunnels and underground openings in cohesive material. *Géotechnique* **30**, 397–416.

Desai, C. S. 1980. A general basis for yield, failure and potential functions in plasticity. *Int. J. Numer. Analyt. Meth. Geomech.* **4**, 361–75.

Drucker, D. C. and W. Prager 1952. Soil mechanics and plastic analysis or limit design. *Q. Appl. Math.* **10**, 157–65.

Eissa, E. A. 1980. *Stress analysis of underground excavations in isotropic and stratified rock using the boundary element method.* PhD thesis, University of London.

Elliott, G. M. and E. T. Brown 1985. Yield of a soft, high porosity rock. *Géotechnique* **35**, 413–23.

Endersbee, L. A. and E. O. Hofto 1963. Civil engineering design and studies in rock mechanics for Poatina underground power station, Tasmania. *J. Instn Engrs, Aust.* **35**, 187–209.

Galileo, G. 1638. *Discorsi e dimostrazioni matematiche, intorno a due nuove scienze attenti alla mechanica e i movimenti locali.* Leyden. See also *Dialogues concerning two new sciences.* Trans. H. Crew and A. de Salvio. New York: Dover, 1952.

Gerrard, C. M. 1977. Background to mathematical modelling in geomechanics. The roles of fabric and stress history. In *Finite elements in geomechanics*, G. Gudehus (ed.), 33–120. London: Wiley.

Gibson, R. E. 1974. Fourteenth Rankine Lecture. The analytical method in soil mechanics. *Géotechnique* **24**, 115–40.

Goodman, R. E. and J. W. Bray 1977. Toppling of rock slopes. In *Rock engineering for foundations and slopes* **2**, 201–34. New York: American Society of Civil Engineers.

Goodman, R. E. and Gen hua Shi 1985. *Block theory and its application to engineering.* Englewood Cliffs, NJ: Prentice-Hall.

Goodman, R. E. and C. M. St John 1977. Finite element analysis for discontinuous rocks. In *Numerical methods in geotechnical engineering*, C. S. Desai and J. T. Christian (eds), 148–75. New York: McGraw-Hill.

Hardy, M. P., C. M. St John and G. Hocking 1979. Numerical modeling of the geomechanical response of a rock mass to a radioactive waste repository. *Proc. 4th Congr.*, *Int. Soc. Rock Mech.*, Montreux, 161–8. Rotterdam: Balkema.

Heyman, J. 1972. *Coulomb's memoir on statics*. Cambridge: Cambridge University Press.

Hill, R. 1950. *The mathematical theory of plasticity*. Oxford: Oxford University Press.

Hocking, G., E. T. Brown and J. O. Watson 1976. Three dimensional elastic stress analysis of underground openings by the boundary integral equation method. *Proc. 3rd symp. applns solid mechs*, Toronto, 203–16. Toronto: University of Toronto Press.

Hoek, E. 1967. A photoelastic technique for the determination of potential fracture zones in rock structures. In *Failure and breakage of rock*, Proc. 8th symp. rock mech., Minneapolis, C. Fairhurst (ed.), 94–112. New York: Society of Mining Engineers, American Institute of Mining, Metallurgical and Petroleum Engineers.

Hoek, E. 1983. Twenty-third Rankine Lecture. Strength of jointed rock masses. *Géotechnique* **33**, 185–222.

Hoek, E. and J. W. Bray 1981. *Rock slope engineering*, 3rd edn. London: Institution of Mining and Metallurgy.

Hoek, E. and E. T. Brown 1980. *Underground excavations in rock*. London: Institution of Mining and Metallurgy.

Jaeger, J. C. and N. G. W. Cook 1979. *Fundamentals of rock mechanics*, 3rd edn. London: Chapman & Hall.

Lambe, T. W. 1973. Thirteenth Rankine Lecture. Predictions in soil engineering. *Géotechnique* **23**, 149–202.

Lemos, J. V., R. D. Hart and P. A. Cundall 1985. A generalized distinct element program for modelling jointed rock mass. In *Fundamentals of rock joints*, O. Stephansson (ed.), 335–43. Luleå: Centak Publishers.

Love, E. A. H. 1927. *A treatise on the mathematical theory of elasticity*, 4th edn. Cambridge: Cambridge University Press.

Maier, G. and T. Hueckel 1979. Nonassociated and coupled flow rules of elasto-plasticity for rock-like materials. *Int. J. Rock Mech. Min. Sci.* **16**, 77–92.

Michelis, P. and E. T. Brown 1986. A yield equation for rock. *Can. Geotech. J.* **23**, 9–17.

Mühlhaus, H-B. 1985. Lower bound solutions for circular tunnels in two and three dimensions. *Rock Mech. Rock Engng* **18**, 37–52.

Müller-Salzburg, L. and Xiu run Ge 1983. Untersuchungen zum mechanischen verhalten geklüfteten gebirges unter Wechsellasten. *Proc. 5th Int. Congr. Rock Mech.*, Melbourne, **1**, A43–9. Rotterdam: Balkema.

Paterson, M. S. 1978. *Experimental rock deformation – the brittle field*. Berlin: Springer.

Poulos, H. G. and E. H. Davis 1974. *Elastic solutions for soil and rock mechanics*. New York: Wiley.

Priest, S. D. and E. T. Brown 1983. Probabilistic stability analysis of variable rock slopes. *Trans Instn Min. Metall.* **92**, A1–12.

Rudnicki, J. W. and J. R. Rice 1975. Conditions for localization of deformation in pressure-sensitive dilatant materials. *J. Mech. Phys. Solids* **23**, 371–94.

Saint-Venant, B. de 1970. Mémoire sur l'establissement des équations différentielles des mouvements intérieurs opérés dans les corps solides ductiles au delà des limites où l'élasticité pourrait les ramener à leur premier élat. *C.r. bedb. Séanc. Acad. Sci, Paris* **70**, 473–80.

Salamon, M. D. G. 1974. Rock mechanics of underground excavations. *Advances in rock mechanics*, Proc. 3rd congr., Int. Soc. Rock Mech., Denver, **1B**, 951–1099. Washington, DC: National Academy of Sciences.

Salamon, M. D. G., J. A. Ryder and W. D. Ortlepp 1964. An analogue solution for determining the elastic response of strata surrounding tabular mine excavations. *J. S. Afr. Instn Min. Metall.* **65**, 115–37.

Stewart, I. J. 1981. *Numerical and physical modelling of underground excavations in discontinuous rock*. PhD thesis, University of London.

Trollope, D. H. 1968. The mechanics of discontinua or clastic mechanics in rock problems. In *Rock mechanics in engineering practice*, O. C. Zienkiewicz and K. G. Stagg (eds), 275–320. London: Wiley.

Tsang, C. F., J. Noorishad and J. S. Y. Wang 1983. A study of coupled thermo-mechanical, thermohydrological and hydromechanical processes associated with a nuclear waste repository in a fractured rock medium. In *Scientific basis for nuclear waste management*, D. G. Brookins (ed.), **6**, 515–22. New York: North Holland.

Vardoulakis, I. 1983. Rigid granular plasticity model and bifurcation in the triaxial test. *Acta Mechanica* **49**, 57–79.

Vardoulakis, I. 1984. Rock bursting as a surface instability problem. *Int. J. Rock Mech. Min. Sci.* **21**, 137–44.

Voegele, M., C. Fairhurst and P. Cundall 1978. Analysis of tunnel support loads using a large displacement, distinct block model. In *Storage in excavated rock caverns*, M. Bergman (ed.), **2**, 247–52. Oxford: Pergamon.

Watson, J. O. 1979. Advanced implementation of the boundary element method for two- and three-dimensionaal elastostatics. In *Developments in boundary element methods – 1*, P. K. Banerjee and R. Butterfield (eds), 31–63. Barking, Essex: Applied Science Publishers.

Watson, J. O. and R. Cowling 1985. Application of three-dimensional boundary element method to modelling of large mining excavations at depth. In *Numerical methods in geomechanics*, Proc. 5th int. symp. numer. methods geomech., Nagoya, T. Kawamoto and Y. Ichikawa (eds), **4**, 1901–10. Rotterdam: Balkema.

Zienkiewicz, O. C. 1977. *The finite element method*, 3rd edn. London: McGraw-Hill.

2 Some applications of elastic theory

J. W. BRAY

2.1 Introduction

In certain circumstances, the rock mass surrounding an underground excavation may behave as an elastic material satisfying Hooke's law, and then linear elastic analysis can be used to make accurate predictions of stress and displacement. More commonly, only part of the structure responds elastically, but the more critically stressed areas exhibit inelastic behaviour by yielding, fracturing, or slipping on surfaces of weakness. Even in these cases elastic analysis is useful in suggesting the influence of the various parameters, and the possible extent of plastic zones. Where applicable, elastic analysis can be used to evaluate a number of factors of importance such as

(a) the maximum and minimum stresses on the boundary of the opening (which is the most critical area);
(b) the boundary displacement induced by the excavation (which allows an assessment to be made of the interaction between supports and the surrounding rock);
(c) the extent of the zone of influence (which allows one to estimate the degree of interaction between neighbouring excavations);
(d) the extent of the overstressed regions (which gives an approximate indication of the areas which actually undergo plastic deformation);
(e) the increase in strain energy, and the dynamic energy released, when an excavation is generated (which is of interest in blasting and rockbursts).

Two topics are discussed in this chapter: (1) the stresses and displacements induced by loads applied to the walls of an opening, and (2) the zone of influence surrounding an elliptical excavation in (a) an isotropic medium and (b) a transversely isotropic medium. In discussing these topics, an attempt is made to introduce points of interest in respect of both the analytical techniques and the practical applications.

32

2.2   Circular holes under internal load

There are various situations where a circular hole is subjected to some form of internal loading and where there is a need for a simple method of calculating the resulting stresses and displacements. An obvious example is in the use of a Goodman borehole jack, where the elastic modulus of a rock is determined by forcing two mutually opposed rams against the sides of a borehole and measuring the consequent change in diameter. Another instance occurs in tunnel design where some assessment has to be made of the stresses and displacements associated with the support forces in order to gauge the interaction between the support system and the surrounding rock. The same question can also arise in the analysis of elastoplastic problems in an infinite medium, where the plastic region is taken to be contained within an imaginary circular boundary and represented by finite elements, and the influence of the surrounding elastic medium is worked out by establishing the relationship between loads and displacements on the circular boundary.

Except for certain elementary cases of loading, the usual way of treating these problems is to employ a complex Fourier analysis (Jaeger & Cook 1964, Duvall & Blake 1967, Goodman *et al.* 1972). The difficulty with this procedure is that it presents the answer in the form of an infinite series, which must be truncated in computation. In certain cases the convergence is rather slow, and the errors could be significant. The method described in this chapter makes repeated use of superposition, and produces answers in a simple form which permits accurate evaluation of stresses and displacements.

2.2.1   *Assumptions, sign convention, and notation*

The mass of rock surrounding the hole is taken to be homogeneous, isotropic, linear, and elastic. It is also assumed to be infinite in extent, with conditions of plane strain prevailing. As regards sign convention, the following are taken to be positive: compressive stresses, and displacement components which result in an increase in coordinates. The most important symbols are as follows:

a = radius of hole

G, ν = shear modulus and Poisson's ratio of rock

u = component of displacement

σ, τ = normal and shear components of stress

F = intensity of line load per unit length of hole

p = intensity of distributed load per unit area

r, θ = polar coordinates, with the pole at the centre of the circle

R, ϕ = polar coordinates, with the pole at some arbitrary point

Angles are measured in radians and are normally to be taken as positive when measured in an anticlockwise sense from a reference axis.

2.2.2 Elementary types of loading

Before proceeding with the general method of analysis, it is worth while considering the following cases of uniform loading which will be referenced at a later stage.

Figure 2.1a shows a circular hole subjected to a uniform radial pressure p. Expressions for stress and displacements are easily obtained from Lamé's theory for thick cylinders by allowing the external diameter to become infinite. An alternative approach is to superpose the stress states given in Figures 2.1b and c. In the first of these a hydrostatic field stress p is imposed in addition to the internal pressure p, and this results in a uniform state of hydrostatic stress p throughout the rock mass. Hence

$$\sigma_r = \sigma_\theta = p, \qquad\qquad \tau_{r\theta} = 0$$
$$u_r = -(1 - 2\nu)pr/2G, \qquad u_\theta = 0$$

(2.1)

In Figure 2.1c, there is no surface traction on the wall of the hole, but there is a hydrostatic field stress of $-p$. The Kirsch equations for this case are

$$\sigma_r = -p(1 - a^2/r^2)$$
$$\sigma_\theta = -p(1 + a^2/r^2)$$
$$u_r = (p/2G)\{(1 - 2\nu)r + a^2/r\}$$
$$\tau_{r\theta} = 0, \qquad u_\theta = 0$$

(2.2)

Adding the expressions given in Equations 2.1 produces

$$\sigma_r = -\sigma_\theta = pa^2/r^2$$
$$u_r = pa^2/2Gr$$
$$\tau_{r\theta} = 0, \qquad u_\theta = 0$$

(2.3)

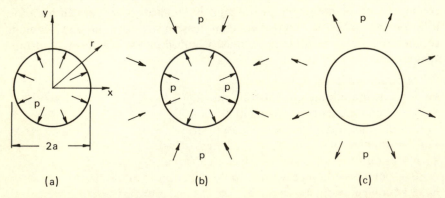

Figure 2.1 A circular hole subjected to three systems of loading: (a) an internal pressure p; (b) an internal pressure p and a hydrostatic field stress p; and (c) a hydrostatic tensile field stress p. By superposition, (a) is equivalent to (b) + (c).

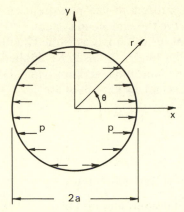

Figure 2.2   A circular hole carrying a horizontal internal load p which is uniformly distributed over the height of the opening.

In Figure 2.2 the wall of the hole is subjected to a horizontal load of intensity p which is uniformly distributed over the height of the opening. Proceeding as in the previous case, one gets the following results.

$$\sigma_r = \frac{p}{2}\,[a^2/r^2 + (4a^2/r^2 - 3a^4/r^4)\cos 2\theta]$$

$$\sigma_\theta = \frac{p}{2}\,[-a^2/r^2 + 3(a^4/r^4)\cos 2\theta]$$

$$u_r = (p/4G)[a^2/r + [4(1 - \nu)a^2/r - a^4/r^3]\cos 2\theta]$$

$$u_\theta = -(p/4G)[2(1 - 2\nu)a^2/r + a^4/r^3]\sin 2\theta$$

(2.4)

The above procedure is applicable only in a few special cases where the load is distributed over the entire surface of the hole. In treating problems of strip or line loading, a different superposition technique must be adopted, and it is this which forms the main theme of the present work. The analysis is developed in three stages, with an increase of generality at each stage, related to the type of loading which can be accommodated:

(1) loading consisting essentially of pairs of equal and opposite forces;
(2) loading comprising any system of balanced forces;
(3) a general system of loading, subject only to the limitations of plane strain.

2.2.3   Force pairs

In Chapter 4 of their book *Theory of elasticity*, Timoshenko and Goodier (1951) show how superposition may be used to solve the problem of a circular disk loaded by two equal and opposite forces applied to its boundary.

Virtually the same procedure will now be used in treating the case of a circular hole subject to the same load system.

The starting point for this analysis is the situation shown in Figure 2.3a, where a tangential line load of intensity F is applied at a point A in the surface of a semi-infinite medium, this medium lying above the plane $y = 0$. The expressions for stress and displacement are

$$\sigma_{R_1} = 2F \sin \phi_1 / \pi R_1$$

$$\sigma_{\phi_1} = \tau_{R\phi_1} = 0$$

$$u_{x_1} = -(F/2\pi G)[2(1 - \nu) \ln R_1 + \tfrac{1}{2} \cos 2\phi_1 + b]$$

$$u_{y_1} = (F/2\pi G)[(1 - 2\nu)\phi_1 + \tfrac{1}{2} \sin 2\phi_1]$$

(2.5)

b is an integration constant, and taking $u_{x_1} \to 0$ as $R_1 \to \infty$ means that b is infinite. For points on the x axis to the right of A, $\phi_1 = \pi/2$ and consequently $u_y = (1 - 2\nu)F/4G$, whereas for points to the left of A, $\phi_1 = -\pi/2$ and $u_y = -(1 - 2\nu)F/4G$. It follows that the surface profile assumes the stepped form shown by broken lines in Figure 2.3a. With a line load F applied at another point B in the opposite direction, the situation is as shown in Figure 2.3b, with a radial state of stress diverging from B, and another stepped surface profile. Equations 2.5 apply, with subscript 1 replaced by 2, and with a positive sign replacing the leading negative sign in the expression for u_{x_1}. When the stresses and displacements for these two cases are combined correctly, we have the result for the two forces F acting simultaneously, as in

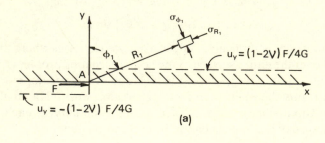

(a)

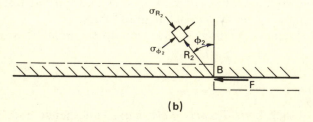

(b)

Figure 2.3 The stress and displacement induced in a semi-infinite medium by horizontal line loads F applied separately at A and B.

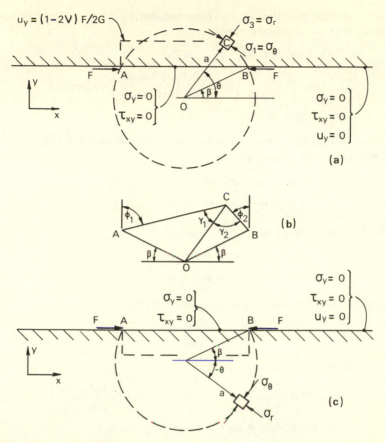

Figure 2.4   (a) The principal stresses and surface displacement produced in a semi-infinite medium by the combined action of two equal and opposite forces F applied at A and B; (b) geometrical features required in the analysis of (a); (c) mirror image of the system in (a).

Figure 2.4a. The constant b in the expression of u_x now vanishes, and the vertical boundary displacement u_y also vanishes to the left of A and to the right of B, whereas over AB it is doubled. If any circle is drawn through points A and B and the state of stress at any point C on this circle is examined, it will be found that the minor principal stress σ_3 passes through the centre O of the circle. The angles of inclination of OB and OC are denoted by β and θ respectively. The angles between AC and OC and between BC and OC are labelled γ_1 and γ_2 respectively. From the geometry of Figure 2.4b it can be shown that

$$\gamma_1 = \phi_2 = \tfrac{1}{2}(\theta + \beta)$$
$$\gamma_2 = \phi_1 = \tfrac{1}{2}(\pi - \theta + \beta)$$
$$R_1 = AC = 2a \cos \phi_2$$
$$R_2 = BC = 2a \cos \phi_1$$

(2.6)

where a = the radius of the circle. Then considering cylindrical components of stress with O as the reference pole,

$$\tau_{r\theta} = \tfrac{1}{2}\sigma_{R_1} \sin 2\gamma_1 - \tfrac{1}{2}\sigma_{R_2} \sin 2\gamma_2$$

$$= \left(\frac{F \sin \phi_1}{2\pi a \cos \phi_2}\right)\sin 2\phi_2 - \left(\frac{F \sin \phi_2}{2\pi a \cos \phi_1}\right)\sin 2\phi_1$$

$$= 0$$

This confirms that the circumferential and radial stresses are the principal components of stress. These are

$$\sigma_1 = \sigma_\theta = \sigma_{R_1} \sin^2 \gamma_1 + \sigma_{R_2} \sin^2 \gamma_2$$

$$\sigma_3 = \sigma_r = \sigma_{R_1} \cos^2 \gamma_1 + \sigma_{R_2} \cos^2 \gamma_2$$

The displacement components add directly.

$$u_x = u_{x_1} + u_{x_2}$$

$$u_y = u_{y_1} + u_{y_2}$$

Substituting from Equations 2.6, we eventually obtain

$$\sigma_\theta = \frac{2F \cos \beta \sin \theta}{\pi a(\sin \theta - \sin \beta)} - \frac{F \cos \beta}{\pi a}$$

$$\sigma_r = F \cos \beta / \pi a$$

$$u_x = (F/2\pi G)\left[\cos \beta \cos \theta + 2(1 - \nu) \ln \left|\frac{\cos \tfrac{1}{2}(\theta + \beta)}{\sin \tfrac{1}{2}(\theta - \beta)}\right|\right] \qquad (2.7)$$

$$u_y = (F/2\pi G)[\cos \beta \sin \theta + (1 - 2\nu)(\beta + \pi/2)]$$

Inverting Figure 2.4a produces the mirror image system shown in Figure 2.4c in which the semi-infinite medium extends downwards from the horizontal surface. Taking the sign convention for β and θ to be the same as before (anticlockwise from x axis positive), all of the Equations 2.7 apply with the exception that the $\pi/2$ term in the expression for u_y now carries a negative sign.

It is now imagined that the semi-infinite media of Figures 2.4a and c are brought into contact and the two horizontal surfaces welded together, except for the part AB. This produces the infinite medium of Figure 2.5a, with line forces $2F$ applied at the end points of the horizontal plane crack AB. The stress and displacement for this situation is still given by Equations 2.7. In understanding the reasoning behind this claim, the following points should be noted:

(a) the equations have been obtained by superposing elastic solutions, and therefore they automatically satisfy the equations of equilibrium, strain-displacement relations, and constitutive equations;

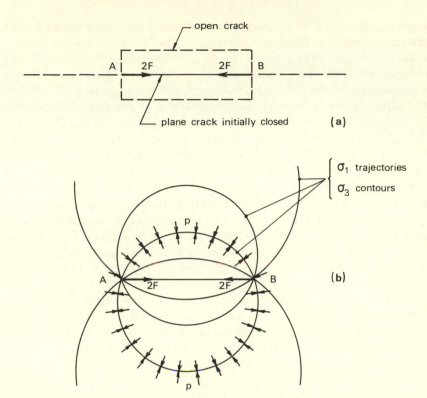

Figure 2.5 Mutually opposed line loads applied to the tips of a plane crack: (a) shows the uniform opening of the crack, (b) shows the σ_1 trajectories which coincide with the σ_3 contours.

(b) the boundary conditions $\sigma_x = \tau_{xy} = 0$ on the surface of the crack are satisfied;

(c) for all points on the plane $y = 0$ outside the crack there is continuity of stress and displacement;

(d) for all points a long way from the crack the stress and displacement tend to zero;

(e) the displacements $u_y = \pm(1 - 2\nu)F/2G$ on AB result in a separation of the crack surfaces, and there is no incompatibility with the physical system.

The peculiar (rectangular) geometry of the deformed ends of the crack is the result of applying concentrated line loads, which lead theoretically to infinite local stresses and strains. It is worth pointing out that if the line loads $2F$ were reversed, the above analysis would no longer be tenable because it would imply an interpenetration of the material lying above and below the crack AB.

The real interest of this work is not to do with the crack, but instead hinges on the circular form of the σ_3 contours and σ_1 trajectories (Fig. 2.5b). This suggests that we could concentrate attention on one of the circles passing through A and B, and imagine that the rock inside the circle was excavated, while at the same time artificial support forces were provided which were identical to those which were originally provided by the mined-out rock. These would maintain the status quo as regards stresses and displacements. The support pressure p would be constant at $\sigma_3 = F \cos \beta / \pi a$ over the whole of the circular boundary, apart from the singular points A and B. To avoid difficulties associated with the peculiar states of stress at A and B the boundary is modified slightly by the insertion of two circular arcs which are designed to bypass the singularities, as in Figure 2.6a. With arcs of finite radius ε, the surface tractions on the arcs vary in a complicated way, but if we allow $\varepsilon \to 0$ then the stress on each arc is influenced only by its own local line load, and it is then easy to prove by integration that the surface force on each arc has an outwardly directed resultant of magnitude F. Combining these with the original line forces leaves inward forces F at A and B, in addition to the pressure p, which is uniformly distributed over the boundary of the circular hole (Fig. 2.6b). The latter can be removed by superposing the effect of a uniform boundary tension of the same magnitude. The relevant expressions are those given in Equations 2.3, with a change of sign. It happens that in most cases of practical interest the applied forces F are directed outwards as in Figure 2.7a, and for this case the previous analysis shows that the boundary values of stress and displacement are

$$\sigma_\theta = \frac{2F \sin \theta \cos \beta}{\pi a (\sin \beta - \sin \theta)}$$

$$u_x = \frac{(1 - \nu)F}{\pi G} \ln \left| \frac{\cos \frac{1}{2}(\theta + \beta)}{\sin \frac{1}{2}(\theta - \beta)} \right| \tag{2.8}$$

$$u_y = -\frac{(1 - 2\nu)F}{2\pi G} (\beta + k\pi/2)$$

where

$$k = \text{sgn} (\sin \theta - \sin \beta)$$

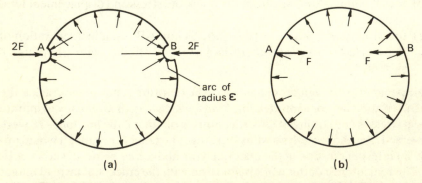

(a)   (b)

Figure 2.6   A circular hole subjected to line loads and distributed loading.

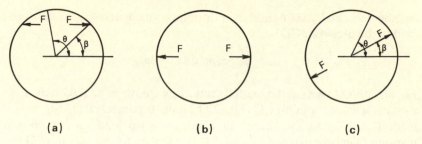

Figure 2.7   A circular hole loaded by (a) a general horizontal force pair, (b) a force pair applied along the horizontal diameter, and (c) a general radial force pair.

These are the results for a horizontal force pair. The case of a radial force pair, where the opposed forces are applied along an arbitrarily inclined diameter, is easily derived. Putting $\beta = 0$ in Equations 2.8 gives the solution for loads applied along the horizontal diameter (Fig. 2.7b). With regard to displacement, it is found that polar components are more useful here than rectangular components, and the change is effected by using the transformation equations

$$u_r = u_x \cos \theta + u_y \sin \theta$$

$$u_\theta = u_y \cos \theta - u_x \sin \theta$$

The effect of rotating the load system through an angle β, as in Figure 2.7c, is obtained by replacing θ by $\theta - \beta$ throughout. The result is

$$\sigma_\theta = -2F/\pi a$$

$$u_r = (F/2\pi G)[2(1 - \nu) \cos (\theta - \beta) \ln |\cot \tfrac{1}{2}(\theta - \beta)|$$
$$- \tfrac{1}{2}(1 - 2\nu)k\pi \sin (\theta - \beta)]$$

$$u_\theta = -(F/2\pi G)[2(1 - \nu) \sin (\theta - \beta) \ln |\cot \tfrac{1}{2}(\theta - \beta)|$$
$$+ \tfrac{1}{2}(1 - 2\nu)k\pi \cos (\theta - \beta)]$$

$$(2.9)$$

where

$$k = \text{sgn} \,[\sin (\theta - \beta)]$$

2.2.4   Distributed loads

In themselves, line loads have no direct practical interest: their value is in providing the means of determining the effect of distributed loads by the process of integration. Two examples of this will now be presented.

Figure 2.8a shows the situation where a uniform radial pressure of intensity p is applied over the arcs $-\alpha < \beta < \alpha$ and $\pi - \alpha < \beta < \pi + \alpha$. Boundary values of stress and displacement are obtained by integrating the expressions in Equations 2.9; the values of these expressions when $F = 1$

will be denoted by σ'_θ, u'_r and u'_θ. For points outside the loaded sections (i.e. within the arcs AB, CD)

$$\sigma_\theta = \int_{-\alpha}^{\alpha} \sigma'_\theta pa \ d\beta = -4p\alpha/\pi$$

For any point Q within the loaded sections, we have to deal with the load on a small element containing Q (Fig. 2.8b) quite separately from the rest of the load. Referring to the theory of a normal strip load on a semi-infinite medium (Timoshenko & Goodier 1951) it will be found that σ_θ at Q due to this elemental load is p. This will be called the local load stress. Hence the total stress at Q is

$$\sigma_\theta = -4p\alpha/\pi + p$$

The displacements are given by

$$u_r = \int_{-\alpha}^{\alpha} u'_r pa \ d\beta$$

$$u_\theta = \int_{-\alpha}^{\alpha} u'_\theta pa \ d\beta$$

In evaluating these integrals, one must take $k = 1$ for that portion of the load corresponding to $\beta < \theta$, and $k = -1$ for that portion where $\beta > \theta$. Summarizing the results:

$$\sigma_\theta = -4p\alpha/\pi + \mu p$$

$$\begin{aligned}
u_r = (1 - \nu)(pa/\pi G)[&\sin (\theta + \alpha) \ln |\cot \tfrac{1}{2}(\theta + \alpha)| \\
&- \sin (\theta - \alpha) \ln |\cot \tfrac{1}{2}(\theta - \alpha)| + 2\alpha \\
&- \tfrac{1}{2}\pi\lambda(1 - 2\nu)/(1 - \nu)]
\end{aligned}$$

$$\begin{aligned}
u_\theta = (1 - \nu)(pa/\pi G)[&\cos (\theta + \alpha) \ln |\cot \tfrac{1}{2}(\theta + \alpha)| \\
&- \cos (\theta - \alpha) \ln |\cot \tfrac{1}{2}(\theta - \alpha)| \\
&+ \ln |\sin (\theta - \alpha)/\sin (\theta + \alpha)| \\
&- \tfrac{1}{2}\pi\zeta \cos \theta \ (1 - 2\nu)/(1 - \nu)]
\end{aligned}$$

(2.10)

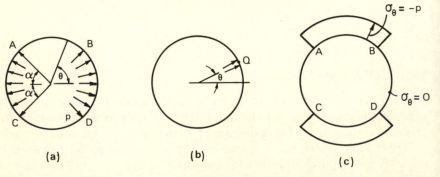

(a)   (b)   (c)

Figure 2.8 (a) Pressure p applied to arcs AC and BD of a circular hole; (b) a load element containing the point Q at which the stress is to be determined; (c) the distribution of circumferential stress along the boundary, when $\alpha = \pi/4$.

where

$$\mu = 1, \qquad \zeta = \sin\theta, \qquad \lambda = 1 - \cos\alpha\cos\theta, \qquad -\alpha < \theta < \alpha$$

$$\mu = 0, \qquad \zeta = \sin\alpha, \qquad \lambda = \sin\alpha|\sin\theta|, \qquad -\pi/2 < \theta < -\alpha, \; \alpha < \theta < \pi/2$$

In the special case where $\alpha = \pi/2$, these equations reduce to $\sigma_\theta = -p$, $u_r = pa/2G$ and $u_\theta = 0$ in agreement with Equations 2.3. The distribution of circumferential stress around the boundary when $\alpha = \pi/4$ is shown in Figure 2.8c. The discontinuous nature of this distribution is to be specially noted. The general problem of normal loading was first solved by Jaeger and Cook (1964), and it is instructive to compare Equations 2.10 with their solution, which is

$$\sigma_\theta = -2p\alpha/\pi + (2p/\pi)\sum_{m=1}^{\infty}(1/m)\sin 2m\alpha\cos 2m\theta$$

$$u_r = (pa/2\pi G)\left\{2\alpha + \sum_{m=1}^{\infty}(1/m)\left[\frac{3-4\nu}{2m-1} + \frac{1}{2m+1}\right]\sin 2m\alpha\cos 2m\theta\right.$$

No expression was given for u_θ. Jaeger and Cook note that the expression for σ_θ can be reduced to the form given in Equation 2.10, and also that the series in the expression for u_r can be summed in the particular cases where $\theta = 0$ and $\theta = \pi/2$. It is in fact possible to carry out the summation for all other values of θ. By way of example, the case where $\alpha < \theta < \pi - \alpha$ will be considered. For simplicity the limits of the various sums will be omitted and it is to be understood that they are from 1 to ∞.

The series is re-cast as follows:

$$\sum\left(\frac{3-4\nu}{2m-1} - \frac{1-2\nu}{m} - \frac{1}{2m+1}\right)(\sin 2m(\theta+\alpha) - \sin 2m(\theta-\alpha))$$

Considering each of the component terms separately,

$$\sum\frac{\sin 2m(\theta+\alpha)}{2m-1} = \cos(\theta+\alpha)\sum\frac{\sin(2m-1)(\theta+\alpha)}{2m-1}$$

$$+ \sin(\theta+\alpha)\sum\frac{\cos(2m-1)(\theta+\alpha)}{2m-1}$$

$$= \frac{\pi}{4}\cos(\theta+\alpha) + \tfrac{1}{2}\sin(\theta+\alpha)\ln\cot\left(\frac{\theta+\alpha}{2}\right)$$

$$\sum\frac{\sin 2m(\theta+\alpha)}{m} = \tfrac{1}{2}[\pi - 2(\theta+\alpha)]$$

$$\sum\frac{\sin 2m(\theta+\alpha)}{2m+1} = \cos(\theta+\alpha)\sum\frac{\sin(2m+1)(\theta+\alpha)}{2m+1}$$

$$- \sin(\theta+\alpha)\sum\frac{\cos(2m+1)(\theta+\alpha)}{2m+1}$$

$$= \frac{\pi}{4}\cos(\theta+\alpha) - \tfrac{1}{2}\sin(\theta+\alpha)\ln\cot\left(\frac{\theta+\alpha}{2}\right)$$

The expressions for sums involving $\sin 2m(\theta - \alpha)$ are obtained by replacing α by $-\alpha$. With these results Jaeger and Cook's expression for u_r reduces to the form given in Equations 2.10. For all those situations where a reduction of this kind can be implemented, the Fourier method of analysis is very useful, but where this is not possible one is faced with the problem of convergence, and the errors associated with the truncation of infinite series. In certain cases only a few terms are required to give an acceptable accuracy, but in other cases many thousands of terms may have to be evaluated.

The second example of distributed loading is that shown in Figure 2.9a where a horizontal pressure of uniform intensity p per unit of height is applied over arcs $-\alpha < \beta < \alpha$ and $\pi - \alpha < \beta < \pi + \alpha$. The boundary stress and the displacement components are given by

$$\sigma_\theta = \int_{-\alpha}^{\alpha} \sigma_\theta' pa \cos \beta \, d\beta, \qquad u_x = \int_{-\alpha}^{\alpha} u_x' pa \cos \beta \, d\beta, \qquad u_y = \int_{-\alpha}^{\alpha} u_y' pa \cos \beta \, d\beta$$

where the primes indicate the expressions given in Equations 2.8 with $F = 1$. On evaluating the integrals it is found that

$$\sigma_\theta = -\frac{2p \sin \theta}{\pi} \left(2\alpha \sin \theta + \cos \theta \ln \left| \frac{\sin (\theta + \alpha)}{\sin (\alpha - \theta)} \right| \right) + kp \cos^2 \theta$$

$$u_x = 2(1 - \nu^2) \frac{pa}{\pi E} \left(\sin \alpha \ln \left| \frac{\cos \theta + \cos \alpha}{\cos \theta - \cos \alpha} \right| - \sin \theta \ln \left| \frac{\sin (\theta + \alpha)}{\sin (\theta - \alpha)} \right| + 2\alpha \cos \theta \right)$$

$$\tag{2.11}$$

$$u_y = -(1 - 2\nu) \frac{pa}{2G} \sin \zeta$$

where

$$k = 0, \qquad \zeta = -\alpha, \qquad (-\pi/2 < \theta < -\alpha)$$

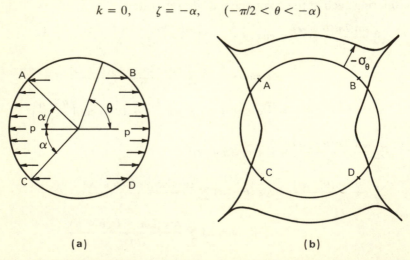

(a)   (b)

Figure 2.9 (a) A horizontal load of intensity p per unit height applied to arcs AC and BD of a circular hole; (b) the distribution of circumferential stress at the boundary.

$$k = 1, \quad \zeta = \theta, \quad (-\alpha < \theta < \alpha)$$
$$k = 0, \quad \zeta = \alpha, \quad (\alpha < \theta < \pi/2)$$

Transforming to cylindrical components of displacement in the normal way,

$$
u_r = \frac{pa}{2\pi G} \left\{ (1 - v) \left[2 \sin \alpha \cos \theta \ln \left| \frac{\cos \theta + \cos \alpha}{\cos \theta - \cos \alpha} \right| \right. \right.
$$

$$
\left. + \sin 2\theta \ln \left| \frac{\sin (\theta - \alpha)}{\sin (\theta + \alpha)} \right| + 4\alpha \cos^2 \theta \right]
$$

$$
\left. - (1 - 2v)\pi \sin \theta \sin \zeta \right\}
\tag{2.12}
$$

$$
u_\theta = \frac{-pa}{2\pi G} \left\{ (1 - v) \left[2 \sin \alpha \sin \theta \ln \left| \frac{\cos \theta + \cos \alpha}{\cos \theta - \cos \alpha} \right| \right. \right.
$$

$$
\left. + 2 \sin^2 \theta \ln \left| \frac{\sin (\theta - \alpha)}{\sin (\theta + \alpha)} \right| + 2\alpha \sin 2\theta \right]
$$

$$
\left. + (1 - 2v)\pi \cos \theta \sin \zeta \right\}
$$

As before, the special case where $\alpha = \pi/2$ provides a check. The above equations reduce to

$$\sigma_\theta = \tfrac{1}{2}p(3 \cos 2\theta - 1)$$
$$u_r = (pa/4G)[1 + (3 - 4v) \cos 2\theta]$$
$$u_\theta = -(pa/4G)[(3 - 4v) \sin 2\theta]$$

and these agree with Equations 2.4.

Returning to the general case ($\alpha \neq \pi/2$), this problem was originally analysed by Goodman *et al.* (1972) in order to predict the performance of their borehole jack. Like Jaeger and Cook, they used complex Fourier analysis, and, when suitably modified to agree with the notation and sign convention of this chapter, their results take the form

$$
\sigma_\theta = \frac{p}{\pi} \left\{ \sum_{m=0}^{\infty} \frac{\sin 2(m + 1)\alpha}{m + 1} [2 \cos 2(m + 1)\theta - \cos 2m\theta + 3 \cos 2(m + 2)\theta] \right.
$$

$$
\left. - 2\alpha + 6\alpha \cos 2\theta \right\}
$$

$$
u_r = \frac{pa}{\pi G} \left\{ \sum_{m=1}^{\infty} \frac{\sin 2m\alpha}{m} \left[\frac{3 - 4v}{2m + 1} \cos 2(m + 1)\theta \right. \right.
$$

$$
\left. + \left(\frac{3 - 4v}{2m - 1} + \frac{1}{2m + 1} \right) \cos 2m\theta + \frac{\cos 2(m - 1)\theta}{2m - 1} \right]
$$

$$
\left. + 2\alpha[1 + (3 - 4v) \cos 2\theta] \right\}
$$

As in the radial pressure example, the infinite series can be summed, and when this is done the above expressions reduce to those given in Equations 2.11 and 2.12.

In applying the theory to the borehole jack, it is reasonable to assume that the displacement of each bearing plate is equal to the horizontal displacement, averaged out over the height of the contact surface, i.e.

$$\bar{u}_x = \frac{1}{2a \sin \alpha} \int_{-a \sin \alpha}^{a \sin \alpha} u_x \, dy = \frac{pa}{E} K(\nu, \alpha)$$

where

$$K(\nu, \alpha) = \frac{2(1 - \nu^2)}{\pi} \left(\frac{\alpha^2}{\sin \alpha} + 2\alpha \cos \alpha - 2 \sin \alpha \ln |\sin \alpha| \right)$$

In the practical instrument $\alpha = \pi/4$ and so

$$K(\nu, \alpha) = 1.5745(1 - \nu^2)$$

The distribution of circumferential stress in this case has the form shown in Figure 2.9b. It is to be noted that this stress becomes theoretically infinite at the edges of the bearing plates ($\theta = \pm\alpha$). It follows that the assumption of uniform horizontal loading is untenable: a different load distribution must be developed in practice, either elastically or as a result of localized failure. Theoretical calculations show that the horizontal displacement is fairly sensitive to the way in which the load is distributed. It may be of interest to note that the vertical displacement of the unloaded parts of the borehole depends only on the magnitude of the load F and not on the way in which it is distributed ($u_y = \pm(1 - 2\nu)F/2G$). Although smaller in magnitude, this measurement may be more reliable for modulus determination. The form of the stress distribution in Figures 2.8c and 2.9b are of special interest. They exemplify two general principles, which apply to other shapes of boundary:

(a) Where there is a discontinuity in the intensity of normal traction on the boundary, there will be a corresponding discontinuity in the value of the tangential stress.
(b) Where there is a discontinuity in the intensity of shear traction on the boundary, the tangential stress becomes infinite.

2.2.5   Balanced load systems

In the previous section the basic load consisted of two equal and opposite forces, and this was referred to simply as a force pair. What is proposed now is to show that any system of boundary forces in mutual equilibrium can be replaced by an equivalent system of force pairs. Once this equivalence is established, one can treat any balanced load system by the procedures developed in the previous section.

By way of illustration we consider a load system made up of three line forces, F_1, F_2, and F_3, applied to boundary points C_1, C_2, and C_3 respectively, as in Figure 2.10a. Points A and B are established at the opposite ends of the horizontal diameter (Fig. 2.10b). Force F_1 is resolved into two components, P_1 and Q_1, these being parallel to the lines AC_1 and BC_1 respectively. Forces F_2 and F_3 are treated in the same manner. Corresponding to each of the forces P_1, P_2, and P_3, equal and opposite forces are now imposed at A (Fig. 2.10c). Likewise, extra forces Q_1, Q_2, and Q_3 are placed at B in opposition to the Q components at C_1, C_2, and C_3. By this procedure, pairs of forces P_1, P_2, P_3, Q_1, Q_2, and Q_3 have been established. However, the introduction of extra forces at A and B has changed the system, and to put matters right, neutralizing forces must be applied at A and B, as shown in Figure 2.10d. Now, the original forces F were in equilibrium and so are the various force pairs P and Q. It follows that the final forces introduced in Figure 2.10d must also be in equilibrium, and this must mean that their resultants are equal in magnitude and opposite in direction. These resultants are labelled H in Figure 2.10e, and since H is the resultant of all the

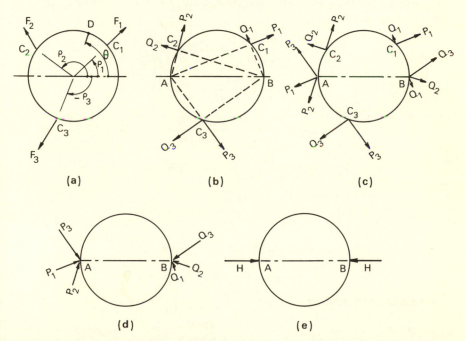

Figure 2.10 (a) A typical system of balanced line loads F applied to the boundary of a circular hole; (b) resolution of each load into a P component passing through A, and a Q component passing through B; (c) force pairs produced by combining the component forces established in (b) with extra forces at A and B; (d) extra negative forces applied at A and B which negate the action of those introduced in (c); (e) force pair H obtained as the resultant of the negative forces introduced in (d).

P forces, or of all the Q forces, then

$$H = \sum_1^3 P \cos (\rho/2) = \sum_1^3 Q \sin (\rho/2)$$

The above reasoning has therefore shown that the original system of forces is equivalent to the six force pairs made up of P and Q forces, plus the extra force pair H. Obviously this type of argument applies irrespective of the number of forces in the original system.

The effect of the P, Q, and H force pairs will be considered separately, and the results summed to produce general expressions for stress and displacement on the boundary. It is necessary to consider two loading configurations, dependent on the angle ρ which defines the position of the load point C: $0 < \rho < \pi$ as in Figure 2.11a and $-\pi < \rho < 0$ as in Figure 2.11b. D is the point at which stress and displacement are to be determined, and the associated angle θ is taken to lie within the range $-\pi < \theta < \pi$.

In applying Equations 2.8 to force pair P, the variables F, β, and θ are replaced by P, $\rho/2$, and $(\theta - \rho/2)$ respectively. This is correct for both Figures 2.11a and b and hence the results are true for the whole load range $-\pi < \rho < \pi$.

$$\sigma_\theta = \frac{2P}{\pi a} \frac{\sin (\theta - \rho/2) \cos \rho/2}{[\sin \rho/2 - \sin (\theta - \rho/2)]}$$

$$= -\frac{P}{\pi a} \left(\frac{\sin \theta/2}{\sin \psi/2} + \frac{\cos \psi/2}{\cos \theta/2} \right)$$

$$u_l = \frac{(1 - \nu)P}{\pi G} \ln \left| \frac{\cos \frac{1}{2}\theta}{\sin \frac{1}{2}\psi} \right|$$

$$u_m = -\frac{(1 - 2\nu)P}{4\pi G} (\rho + k_1 \pi)$$

where

$$\psi = \theta - \rho$$

$$\begin{aligned} k_1 &= \text{sgn} \left[\sin (\theta - \rho/2) - \sin \tfrac{1}{2}\rho \right] \\ &= \text{sgn} (\sin \tfrac{1}{2}\psi \cos \tfrac{1}{2}\theta) \\ &= \text{sgn} (\psi), \quad \text{since } \cos \theta/2 > 0 \end{aligned}$$

Note that, by definition,

$$\text{sgn} (x) = \begin{cases} 1 & \text{when } x > 0 \\ 0 & \text{when } x = 0 \\ -1 & \text{when } x < 0 \end{cases}$$

Considering force pair Q, with the configuration of Figure 2.11a, F, β, and θ must be replaced by Q, $(\pi/2 - \rho/2)$, and $(\theta + \pi/2 - \rho/2)$ respectively.

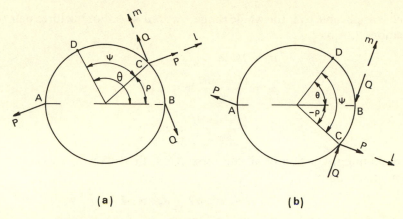

Figure 2.11   The geometry associated with two positions of the load point C: (a) $0 < \rho < \pi$ and (b) $-\pi < \rho < 0$.

Then

$$\sigma_\theta = \frac{2Q}{\pi a} \frac{\cos (\theta - \rho/2) \sin \rho/2}{[\cos \rho/2 - \cos (\theta - \rho/2)]}$$

$$= \frac{Q}{\pi a} \left(\frac{\cos \frac{1}{2}\theta}{\sin \frac{1}{2}\psi} - \frac{\cos \frac{1}{2}\psi}{\sin \frac{1}{2}\theta} \right)$$

$$u_l = \frac{(1 - 2\nu)Q}{4\pi G} (\rho + k_2 \pi)$$

$$u_m = \frac{(1 - \nu)Q}{\pi G} \ln \left| \frac{\sin \frac{1}{2}\theta}{\sin \frac{1}{2}(\theta - \rho)} \right|$$

where

$$k_2 = \text{sgn } [\cos \rho/2 - \cos (\theta - \rho/2)] - 1$$

$$= \text{sgn } (\sin \theta/2 \sin \tfrac{1}{2}\psi) - 1$$

$$= \text{sgn } (\theta) \, \text{sgn } (\psi) - 1$$

Changing to the configuration of Figure 2.11b, F, β, and θ must be replaced by $-Q$, $(\pi/2 + \rho/2)$, and $(\theta + \pi/2 - \rho/2)$ respectively. It will be found that the formulae are the same as before, except for k_2 which is reversed in sign, i.e.

$$k_2 = 1 - \text{sgn } (\theta) \, \text{sgn } (\psi)$$

Fortunately, the two expressions for k_2 can be encompassed in the one formula,

$$k_2 = \text{sgn } (\psi) - \text{sgn } (\theta)$$

which is applicable over the whole range $-\pi < \rho < \pi$. For the force pair H, Equations 2.8 give

$$\sigma_\theta = 2H/\pi a$$

$$u_x = \frac{(1 - \nu)H}{\pi G} \ln |\tan \tfrac{1}{2}\theta|$$

$$u_y = \frac{(1 - 2\nu)H}{4\pi G} \pi \, \text{sgn} \, (\theta)$$

In combining the action of all the force pairs, the following relations are taken into account.

$$u_x = u_m \cos \rho/2 - u_n \sin \rho/2$$

$$u_y = u_m \sin \rho/2 + u_n \cos \rho/2$$

$$F_x = P \cos \rho/2 - Q \sin \rho/2$$

$$F_y = P \sin \rho/2 + Q \cos \rho/2$$

$$H = \sum P \cos \rho/2 = \sum Q \sin \rho/2$$

$$\sum P \sin \rho/2 = \sum Q \cos \rho/2 = 0$$

$$\sum F_x = \sum F_y = 0$$

After some reduction, the boundary stress and displacement are found to be

$$\sigma_\theta = \frac{1}{\pi a} \sum [F_y \cos \tfrac{1}{2}(\theta + \rho) - F_x \sin \tfrac{1}{2}(\theta + \rho)] \, \text{cosec} \, \psi/2$$

$$u_x = \frac{1}{4\pi G} \sum \{(1 - 2\nu)F_y[\rho + \pi \, \text{sgn} \, (\psi)] - 4(1 - \nu)F_x \ln |\sin \psi/2|\} \quad (2.13)$$

$$u_y = -\frac{1}{4\pi G} \sum \{(1 - 2\nu)F_x[\rho + \pi \, \text{sgn} \, (\psi)] + 4(1 - \nu)F_y \ln |\sin \psi/2|\}$$

where $\psi = \theta - \rho$.

To extend the analysis so that stresses can be evaluated at points outside the boundary, it is necessary to return to Figure 2.10 and apply principles already established regarding the equivalence of load systems. It should be seen that the following load systems are equivalent to each other in the sense that they produce the same distribution of stress outside the circular boundary.

(a) Line loads F_1, F_2, and F_3 applied to the boundary of the hole.
(b) Force pairs $P_1, P_2, P_3, Q_1, Q_2, Q_3$, and H applied to the boundary of the hole.
(c) Force pairs $P_1, P_2, P_3, Q_1, Q_2, Q_3$, and H applied to semi-infinite media, plus a uniform pressure of p applied to the boundary of the hole, where

$$p = \frac{1}{\pi a} \{\sum (P \cos \beta/2 + Q \sin \beta/2) - H\}$$

$$= \frac{1}{\pi a} \sum P \cos \beta/2 = \frac{1}{2\pi a} \sum F_r$$

where F_r = radial component of F.

(d) The original system of line loads F_1, F_2, and F_3 applied to semi-infinite media, together with a uniform pressure p applied to the boundary of the hole.

Calculations of stresses for the actual load system (a) are carried out using the equivalent system (d). There is no need to make any special assumption regarding the orientation of the plane boundary in the semi-infinite calculations, since the expression for the radial stress σ_R (as given in Eqn 2.5) is the same for all orientations.

Before proceeding with the general case, the method will be checked by finding the boundary stress at point D in Figure 2.10a. For convenience each of the forces F is resolved into its radial and tangential components, F_r and F_t respectively. Noting that distance CD $= R = 2a \sin \frac{1}{2}\psi$

$$\sigma_\theta = \sum \sigma_R \cos^2 \frac{1}{2}\psi - \frac{1}{2\pi a} \sum F_r$$

$$= \sum (F_t \cos \frac{1}{2}\psi - F_r \sin \frac{1}{2}\psi) \frac{\cos^2 \frac{1}{2}\psi}{\pi a \sin \frac{1}{2}\psi} - \frac{1}{2\pi a} \sum F_r$$

$$= \frac{1}{\pi a} \sum \{F_t \cot \frac{1}{2}\psi - F_r - \frac{1}{2}[F_r \cos \psi + F_t \sin \psi]\}$$

Thus

$$\sigma_\theta = \frac{1}{\pi a} \sum (F_t \cot \frac{1}{2}\psi - F_r) \qquad (2.14)$$

Note that the square bracketed term vanishes because the sum of such terms equals the resultant force in the direction of OD, and for a balanced force system this must be zero.

Expressing F_t and F_r in terms of F_x and F_y, Equation 2.14 reverts to its previous form in Equation 2.13.

Attention is now transferred to the general point D with coordinates x, y (Fig. 2.12). F is a typical line load applied at point C on the boundary. The line CD has a length R and inclination ξ, where

$$R = [(x - a \cos \rho)^2 + (y - a \sin \rho)^2]^{1/2}$$

$$\xi = \tan^{-1} \left(\frac{y - a \sin \rho}{x - a \cos \rho}\right)$$

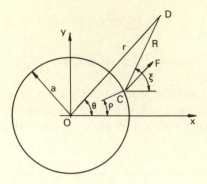

Figure 2.12   The relevant geometrical factors used in analysing the stress at the general point D.

The radial stress at D is

$$\sigma_R = \frac{2}{\pi R}\,(F_x \cos \xi + F_y \sin \xi)$$

with rectangular components

$$\sigma_x' = \sigma_R \cos^2 \xi$$

$$\sigma_y' = \sigma_R \sin^2 \xi$$

$$\tau_{xy}' = \sigma_R \sin \xi \cos \xi$$

Taking all line loads into account and allowing for an internal pressure $p = 1/\pi a\,\Sigma\,F_r$,

$$\sigma_x = \sum \sigma_x' + p\,\frac{a^2}{r^2}\cos 2\theta$$

$$\sigma_y = \sum \sigma_y' - p\,\frac{a^2}{r^2}\cos 2\theta$$

$$\tau_{xy} = \sum \tau_{xy}' + p\,\frac{a^2}{r^2}\sin 2\theta$$

where $r = OD$, and $\theta =$ the inclination of OD.

An example is now chosen to illustrate the application of Equation 2.14. In lining a pressure tunnel it is recognized that the most difficult area to grout is in the region of the crown, and unless care is taken the support provided by the lining may be significantly reduced in this area. An extreme case is modelled in Figure 2.13a, where all contact is lost over the arc ABC which subtends an angle of $2\pi/3$ at the centre O. Over the remainder of the wall, the normal and shear components of traction, p and q, are taken to vary as follows

$$p = p_0 \cos (3\rho/4)$$

$$q = (10p_0/7\sqrt{3}) \sin (3\rho/2)$$

where p_0 = maximum support pressure at lowest point L, ρ = angle measured from L $(-\pi < \rho < \pi)$. The values of the coefficients are chosen to make p and q vanish at A and C, and also to ensure that the resultant load is zero. In Equation 2.14, replacing F_t by $qa\,d\rho$ and F_r by $pa\,d\rho$, and integrating over the loaded sector,

$$\sigma_\theta = -\frac{p_0}{\pi} \int_{-2\pi/3}^{2\pi/3} \left[\frac{10}{7\sqrt{3}} \sin \frac{3\rho}{2} \cot \tfrac{1}{2}\psi + \cos (3\rho/4)\right] d\rho + \frac{p}{2}\left(1 + \frac{\cos\theta + \tfrac{1}{2}}{|\cos\theta + \tfrac{1}{2}|}\right)$$

where the last term is the local load stress. This reduces to

$$\sigma_\theta = -\frac{20p_0}{7\sqrt{3}\pi}\left[\sin\frac{3\theta}{2}\ln\left|\frac{\tan(\theta/4 - \pi/6)}{\tan(\theta/4 + \pi/6)}\right| + 2\sqrt{3}\cos\theta\right] - \frac{8p_0}{3\pi}$$

$$+ \tfrac{1}{2}p_0 \cos (3\theta/4)\left(1 + \frac{\cos\theta + 1/2}{|\cos\theta + 1/2|}\right)$$

where $-\pi < \theta < \pi$. Figure 2.13b shows how this stress is distributed around the boundary. The distribution curve is continuous at all points due to the fact that there are no discontinuities in the distribution of shear and normal surface tractions on the hole boundary. It is instructive to compare the present situation with that illustrated in Figures 2.8c and 2.9b.

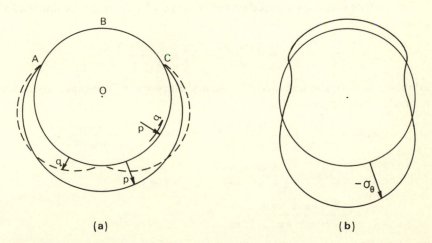

(a)   (b)

Figure 2.13 Pressure tunnel example: (a) the assumed distributions of normal and tangential surface tractions; (b) the calculated distribution of circumferential stress at the wall of the tunnel.

2.2.6 General system of loading

Up to this point, all the load systems applied to the boundary of the hole have been balanced, that is they have constituted a set of forces in equilibrium. This restriction will now be removed, but that does not imply that the system as a whole will not be in equilibrium: it means simply that the boundary forces have a resultant, and this resultant is balanced by forces applied elsewhere in the medium. For instance, with a tensioned rockbolt the force applied by the bolt head to the wall of the hole is balanced by the force applied at the anchor point. With the idealized infinite model that is used in most analyses, it is often necessary to assume that the boundary loads are balanced by reactive forces at infinity.

Any system of 'unbalanced' boundary forces can be resolved into a set of normal and tangential components. Hence the analysis of any general problem depends on knowing the response to an isolated **normal load** and an isolated **tangential load**, and an investigation will now be made of these special cases.

Figure 2.14a shows a line load N applied to point A on the wall of a hole, with the reactive force N distributed over a circle of infinite radius. To render this problem tractable by the previous methods, it is necessary to apply to the boundary extra surface tractions which have a resultant of $-N$ (Fig. 2.14b). These tractions are taken to be those produced on the circle by a negative line load N applied to point O in an infinite medium. Having solved for the combined action of the forces in Figures 2.14a and b, it is necessary to remove the effect of the extra tractions by subtracting the stresses due to the line load of Figure 2.14b.

The surface tractions induced on the boundary by the negative line load at O are

$$\sigma_r = -\frac{3 - 2\nu}{1 - \nu}\frac{N \cos \rho}{4\pi a}, \qquad \tau_{r\rho} = \frac{1 - 2\nu}{1 - \nu}\frac{N \sin \rho}{4\pi a}$$

Allowing also for the line load N at A, and employing Equation 2.14 for the boundary stress at any point D,

$$\sigma_\theta = -\frac{1}{\pi a}\int_{-\pi}^{\pi}\left(\frac{1 - 2\nu}{1 - \nu}\frac{N \sin \rho}{4\pi a}\cot \tfrac{1}{2}\psi - \frac{3 - 2\nu}{1 - \nu}\frac{N \cos \rho}{4\pi a}\right)a\, d\rho$$

$$- N/\pi a - \frac{3 - 2\nu}{1 - \nu}\frac{N \cos \theta}{4\pi a}$$

$$= -\frac{5 - 6\nu}{1 - \nu}\frac{N \cos \theta}{4\pi a} - \frac{N}{\pi a}$$

The $[(3 - 2\nu)/(1 - \nu)](N \cos \theta/4\pi a)$ term is the stress due to the local intensity of the distributed load, as discussed in a previous section.

The line load at O produces a circumferential stress at D of magnitude

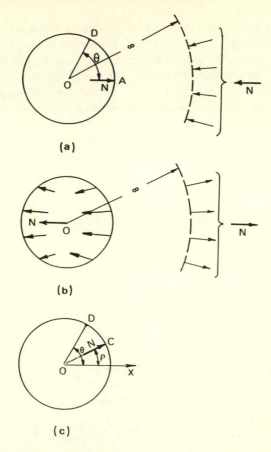

Figure 2.14 (a) A horizontal line load N applied at point A of a circular boundary, with a reactive force of $-N$ at infinity; (b) a horizontal line load $-N$ applied at point O of an infinite medium, generating tractions on the circular surface having a resultant of $-N$; (c) normal line load – general case.

$[(1 - 2\nu)/(1 - \nu)](N \cos \theta/4\pi a)$. Subtracting this from the previous value of σ_θ gives

$$\sigma_\theta = -\frac{N}{2\pi a} \left(\frac{3 - 4\nu}{1 - \nu} \cos \theta + 2 \right)$$

In the general case when N is applied at angle ρ as in Figure 2.14c

$$\sigma_\theta = -\frac{N}{2\pi a} \left(\frac{3 - 4\nu}{1 - \nu} \cos \psi + 2 \right)$$

where $\psi = \theta - \rho$.

In treating the problem of an isolated tangential load T, the procedure is very similar to that used above. Extra surface tractions are applied to the boundary which, together with T, produce a balanced system of loads. The effect of the extra tractions is then eliminated by simple subtraction.

Figure 2.15a shows the problem to be solved. Figure 2.15b shows the same case with the addition of a normal force $N(=T)$, which has been introduced to produce a zero resultant force. At first it was thought that any balanced load system employed in the analysis must also have a zero moment. This proves to be a superfluous requirement, and the load system of Figure 2.15b is taken to be balanced even though it constitutes a couple of moment Ta.

Applying Equation 2.14, the boundary stress is

$$\sigma_\theta = \frac{T}{\pi a} (\cot \tfrac{1}{2}\psi - 1)$$

Removing the effect of N as an isolated boundary force,

$$\sigma_\theta = \frac{T}{\pi a} (\cot \tfrac{1}{2}\psi - 1) + \frac{N}{\pi a} \left\{ 1 + \frac{3 - 4\nu}{2(1 - \nu)} \cos \left[\theta - (\rho - \pi/2) \right] \right\}$$

Thus

$$\sigma_\theta = \frac{T}{\pi a} \left[\cot \tfrac{1}{2}\psi - \frac{3 - 4\nu}{2(1 - \nu)} \sin \psi \right] \tag{2.16}$$

For any system of applied loads, the boundary stress is given by summing terms having the form of Equations 2.15 and 2.16. For consistency with previous work, N and T are replaced by F_r and F_t respectively, and with a slight rearrangement the sum becomes

$$\sigma_\theta = \frac{1}{\pi a} \sum (F_t \cot \tfrac{1}{2}\psi - F_r) - \frac{3 - 4\nu}{2(1 - \nu)\pi a} \sum (F_r \cos \psi + F_t \sin \psi) \tag{2.17}$$

The second sum is equal to the component of the resultant load in the direction of OD, and with a balanced load system this resultant vanishes, Equation 2.14 then being regained.

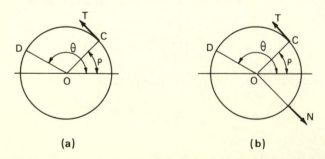

(a)   (b)

Figure 2.15   (a) Tangential line load T applied to a general point on the boundary of a circular hole; (b) tangential line load T balanced by an equal and opposite normal line load N.

It will be noticed that with balanced load systems the stress distribution is independent of the elastic constants, and only when the load applied to the boundary has a non-zero resultant is the stress influenced by the value of Poisson's ratio.

To complete this section of the work, it is intended to derive formulae for the boundary displacement and the stress distribution outside the boundary. This could be done using the procedure just established, but the following approach is to be preferred as it involves less effort.

Referring to Figure 2.17a, the aim is to find the components of stress at any point Q and the displacement at any boundary point D. Initially, attention is focused on the stress at the boundary point D since the solution has to satisfy boundary conditions. The first load system to be considered is that shown in Figure 2.16a where force pairs $2N$ and $2T$ are applied to the surfaces of two semi-infinite media which abut one another along the horizontal plane through O. The stress at D is compounded from a radial stress σ_r acting along OD and a radial stress σ_R acting along AD, where

$$\sigma_r = -(2/\pi a)(N \cos \theta + T \sin \theta)$$

$$\sigma_R = (2/\pi R)(N \cos \phi + T \sin \phi) = (T \cot \theta/2 - N)/\pi a$$

These combine to produce

$$\sigma_r = -[N(1 + 3 \cos \theta) + 3T \sin \theta]/2\pi a$$

$$\sigma_\theta = -[N(1 + \cos \theta) + T \sin \theta - 2T \cot \theta/2]/2\pi a$$

$$\tau_{r\theta} = -[N \sin \theta - T(1 + \cos \theta)]/2\pi a$$

The surface tractions σ_r and $\tau_{r\theta}$ produce a resultant force on the boundary having x, y components of $-N$ and $-T$ respectively. These components can be eliminated by introducing additional forces N and T at the centre O, as in Figure 2.16b, and treating these as line loads applied to an infinite medium.

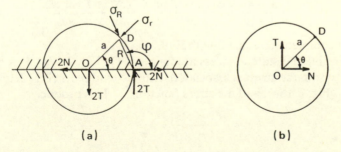

(a)   (b)

Figure 2.16   (a) Force pairs N and T applied to the surfaces of two abutting semi-infinite media producing radial stresses σ_R and σ_r at point D; (b) line loads N and T applied at point O of an infinite medium.

The extra stresses are

$$\sigma_r = \frac{3 - 2\nu}{1 - \nu} \frac{N\cos\theta + T\sin\theta}{4\pi a}$$

$$\sigma_\theta = -\frac{1 - 2\nu}{1 - \nu} \frac{N\cos\theta + T\sin\theta}{4\pi a}$$

$$\tau_{r\theta} = \frac{1 - 2\nu}{1 - \nu} \frac{T\cos\theta - N\sin\theta}{4\pi a}$$

Combining with the previous stresses gives

$$\sigma_r = -\frac{3 - 4\nu}{1 - \nu} \frac{N\cos\theta + T\sin\theta}{4\pi a} - \frac{N}{2\pi a}$$

$$\sigma_\theta = -\frac{3 - 4\nu}{1 - \nu} \frac{N\cos\theta + T\sin\theta}{4\pi a} - \frac{N}{2\pi a} + \frac{T\cot\theta/2}{\pi a}$$

$$\tau_{r\theta} = -\frac{3 - 4\nu}{1 - \nu} \frac{T\cos\theta - N\sin\theta}{4\pi a}$$

The surface tractions can be made to vanish by superposing a further stress distribution derived from the stress function

$$U = (bN/r)\cos\theta + (cT/r)\sin\theta + dN\ln r \qquad (2.18)$$

The associated stress components for $r = a$ are

$$\sigma_r = -2(bN\cos\theta + cT\sin\theta)/a^3 + dN/a^2$$

$$\sigma_\theta = 2(bN\cos\theta + cT\sin\theta)/a^3 - dN/a^2$$

$$\tau_{r\theta} = -2(cT\cos\theta - bN\sin\theta)/a^3$$

Elimination of the surface tractions requires

$$b = -\frac{3 - 4\nu}{1 - \nu} \frac{a^2}{8\pi}, \qquad c = -\frac{3 - 4\nu}{1 - \nu} \frac{a^2}{8\pi}, \qquad d = -\frac{a}{2\pi} \qquad (2.19)$$

Then

$$\sigma_\theta = -\frac{3 - 4\nu}{1 - \nu} \frac{N\cos\theta + T\sin\theta}{2\pi a} - \frac{N}{\pi a} + \frac{T\cot\theta/2}{\pi a}$$

which is in complete agreement with Equation 2.17.

To determine the state of stress at any point Q (Fig. 2.17a), it is merely necessary to sum the stresses associated with the load pairs (Fig. 2.16a), the line forces (Fig. 2.16b), and the stress function U. This produces

$$\sigma_r = \frac{(6\nu - 5)S_1}{4(1 - \nu)\pi} + \frac{2S_2}{\pi} - S_3$$

$$\sigma_\theta = \frac{(2\nu - 1)S_1}{4(1 - \nu)\pi} + \frac{2S_4}{\pi} + S_3 \qquad (2.20)$$

$$\tau_{r\theta} = S_5/\pi + S_6$$

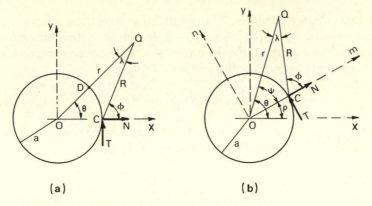

Figure 2.17   Line loads N and T applied at point C on the boundary of a hole, where C is (a) on the horizonal axis and (b) on an axis m inclined at angle ρ.

where

$$S_1 = (N \cos \theta + T \sin \theta)/r$$

$$S_2 = (1/R)(N \cos \phi + T \sin \phi) \cos^2 \lambda$$

$$S_3 = 2(bN \cos \theta + cT \sin \theta)/r^3 - dN/r^2$$

$$S_4 = (1/R)(N \cos \phi + T \sin \phi) \sin^2 \lambda$$

$$S_5 = (1/R)(N \cos \phi + T \sin \phi) \sin 2\lambda$$

$$S_6 = \frac{1 - 2\nu}{1 - \nu} \frac{T \cos \theta - N \sin \theta}{4\pi r} - \frac{2(bN \sin \theta - cT \cos \theta)}{r^3}$$

(2.21)

Where the load point is at some point C defined by the angle ρ (Fig. 2.17b), the above expressions hold with θ replaced by $\psi (= \theta - \rho)$.

2.2.7   Numerical integration

Consider now the case of distributed loading expressed in terms of surface tractions p_r (radial) and p_t (tangential). Equations 2.20 still apply, provided that a suitable adjustment is made to the meaning of the functions S. In Equations 2.21 it is necessary to replace θ by ψ, N by $p_r a \, d\rho$, and T by $p_t a \, d\rho$, and then integrate the resulting expressions. Thus

$$S_1 = \int (p_r \cos \psi + p_t \sin \psi)(1/r)a \, d\rho$$

$$S_2 = \int (1/R)(p_r \cos \phi + p_t \sin \phi) \cos^2 \lambda a \, d\rho, \text{ etc.}$$

It is not always possible to obtain closed-form expressions for these integrals, and it is then necessary to employ a numerical procedure, such as the trapezoidal rule, Simpson's rule, or Gaussian quadrature. It is important to realize that, whatever method is adopted, large errors can be introduced if there are singularities in the integrands, either on the path of integration or close by. To illustrate this difficulty and how it may be overcome, an attempt will now be made to evaluate the following integral

$$I = \int p_r \cos \phi \cos^2 \lambda (a/R) \, d\rho \tag{2.22}$$

which forms part of the function S_2. It will be assumed that $p_r = 1$, $a = 1$, and the limits of the integral are $\pi/3$ and $-\pi/3$. Then

$$I = \int_{-\pi/3}^{\pi/3} g \, d\rho$$

where $g = \cos \phi \cos^2 \lambda/R$. Reference to Figure 2.17b shows that

$$R = (r^2 + a^2 - 2ar \cos \psi)^{1/2}$$

$$\cos \phi = (r^2 - a^2 - R^2)/2aR$$

$$\cos \lambda = (r^2 - a^2 + R^2)/2rR$$

With $p_r = $ constant $(=1)$, the integral can be worked out analytically, giving

$$I = \frac{1}{2\pi} \left[\frac{(r - \cos \psi) \sin \psi}{R^2} + \beta - \frac{\psi}{2} \right]_{-\pi/3}^{\pi/3} \tag{2.23}$$

where

$$\beta = \arctan \left(\frac{r + 1}{r - 1} \tan \frac{\psi}{2} \right)$$

By comparing results with those given by this formula, one can determine the errors associated with the numerical methods. For any arbitrary value of n the value of I given by the trapezoidal rule is

$$I_T = h(\tfrac{1}{2}g_1 + g_2 + g_3 + \cdots + g_{n-1} + \tfrac{1}{2}g_n) \tag{2.24}$$

where $h = 2\pi/3n$, g_1, g_2, g_3, etc. = the values of g when $\rho = -\pi/3$, $-\pi/3 + h$, $-\pi/3 + 2h$, etc. Similarly, Simpson's rule gives

$$I_S = \tfrac{1}{3}h(g_1 + 4g_2 + 2g_3 + \cdots + 4g_{n-1} + g_n) \tag{2.25}$$

To use the Gauss–Legendre integration formula it is necessary to change the variable ρ so that the integration limits are 1 and -1. This is effected by putting $u = 3\rho/\pi$, when

$$I = \frac{\pi}{3} \int_{-1}^{1} \frac{\cos \phi \cos^2 \lambda}{R} \, du$$

According to Gauss's method this integral can be expressed approximately as

$$I_G = \frac{\pi}{3} \sum_{i=1}^{n} A_i g(u_i) \tag{2.26}$$

where the u_i are the zeros of the Legendre polynomials $P_n(u)$, and the A_i are known coefficients. Textbooks on numerical analysis give tabulated values of u_i and A_i for various values of n. In the following computations, n will be taken equal to 6 in the trapezoidal and Gauss methods and 7 when applying Simpson's rule. Using all methods, results are obtained for $r = 1.1$ and for values of θ lying in the range $0 < \theta < 1.4$ radians, and these are displayed in Figure 2.18. It is evident that none of the numerical methods gives satisfactory results; in fact the Gauss method, which is normally far superior to the other two, produces here the worst set of results. For all the methods, peaks are found where θ is close to the particular values of θ at which g is determined. These special values of θ are indicated on a horizontal line below the θ axis, T, S, and G denoting trapezoidal, Simpson, and Gauss points. Broken lines are used to draw attention to the relationship between the special values of θ and the peaks in the trapezoidal case. The reason for this type of behaviour can be found by examining the way the integrand g varies with ψ, as shown in Figure 2.19. For the given value of $r(1.11)$, the curve is very peaky, and as $r \to 1$ the curve tends to assume the form of a Dirac delta function, i.e. infinite at $\psi = 0$ and zero for all other values of ψ. With the 6-point Gaussian formula, the values of u_i are $\pm 0.932\,469\,51$, $\pm 0.661\,209\,39$, and $\pm 0.238\,619\,19$, and in the present problem the corresponding values of ρ are approximately ± 0.9765, ± 0.6924, and ± 0.2499 radians. If $\theta = 0$ then ψ takes on the same set of values as ρ and the location of these is shown in Figure 2.19 by points marked on a horizontal line labelled '$\theta = 0$'. It will be seen that the corresponding ordinates g are all very small and consequently the estimated value of I given by Equation 2.26 is

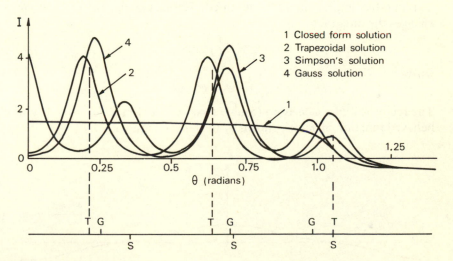

Figure 2.18 Evaluation of the integral I given in Equation 2.22 for a range of values of θ. The results of three numerical methods are compared with the closed form solution.

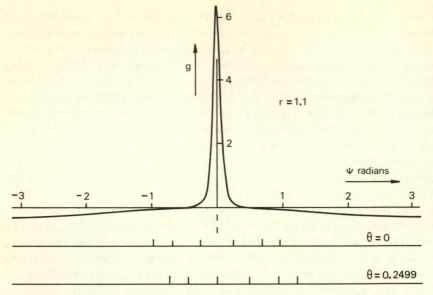

Figure 2.19   The integrand g in Equation 2.22 and its variation with the angle ψ.

also very small. Alternatively, taking $\theta = 0.2499$, the set of ψ values are $-0.7266, -0.4425, 0, 0.4998, 0.9423, 1.2264$, and these are marked on a line labelled '$\theta = 0.2499$' in Figure 2.19. The function $g(u_3)$ is now the peak value and consequently I_G is itself large.

One of the ways out of this difficulty is to look for a suitable change of variable. The factor causing the trouble is R in the integrand, which tends to zero as $r \to a$ and $\psi \to 0$. If integration is taken over ϕ instead of ρ, a considerable improvement is obtained. Replacing $d\rho$ by $-R\, d\phi/r \cos \lambda$ changes the integral to

$$I = -\int g'\, d\phi$$

where

$$g' = \cos \phi \cos \lambda / r$$

The reciprocal of r causes no trouble since $r \lessdot 1$. In fact, g' is a very well behaved function of ϕ, as is shown by the plot in Figure 2.20.

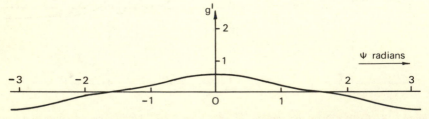

Figure 2.20   The integrand $g' = \cos \phi \cos \lambda / r$ and its variation with angle ψ.

For evaluation by Gaussian quadrature, the integral is changed to

$$I_G = \tfrac{1}{2}(\phi_a - \phi_b) \int_{-1}^{1} g' \, du$$

where

$$u = (2\phi - \phi_a - \phi_b)/(\phi_b - \phi_a)$$
$$\phi_a = \arctan \{[R^2/\sin^2 (\theta + \pi/3) - 1]^{-1/2}\}$$
$$\phi_b = \arctan \{[R^2/\sin^2 (\theta - \pi/3) - 1]^{-1/2}\}$$

The values of I obtained using this procedure are very close to the values given by the closed form solution, so close in fact that plotted at the scale of Figure 2.18 the differences are not discernible. The maximum error is found to be 0.25%.

There remains the question of boundary displacement, and returning to Figure 2.16a the aim is to find the rectangular components of displacement (u_x, u_y) at the typical point D.

Under the action of the force pairs in Figure 2.16a,

$$u_x = -(N/4\pi G)[\cos \theta + \cos 2\theta + 4(1 - \nu) \ln |2 \sin \theta/2|]$$
$$- (T/4\pi G)\{\sin \theta + \sin 2\theta + (1 - 2\nu)[\theta - \pi \, \text{sgn} \, (\theta)]\}$$
$$u_y = -(N/4\pi G)\{\sin \theta + \sin 2\theta - (1 - 2\nu)[\theta - \pi \, \text{sgn} \, (\theta)]\}$$
$$+ (T/4\pi G)[\cos \theta + \cos 2\theta - 4(1 - \nu) \ln |2 \sin \theta/2|]$$

$$(2.27)$$

Under the action of the line loads in Figure 16b,

$$u_x = \{-N[(3 - 4\nu) \ln a + \sin^2 \theta] + \tfrac{1}{2}T \sin 2\theta\}/8\pi(1 - \nu)G$$
$$u_y = \{-T[(3 - 4\nu) \ln a + \cos^2 \theta] + \tfrac{1}{2}N \sin 2\theta\}/8\pi(1 - \nu)G$$

$$(2.28)$$

The displacements associated with the stress function of Equation 2.18 are worked out as follows. Putting $bN \cos \theta + cT \sin \theta = \zeta$ and $CT \cos \theta - bN \sin \theta = \eta$, the strain components become

$$\varepsilon_r = \frac{1 - \nu^2}{E} \left(\sigma_r - \frac{\nu\sigma_\theta}{1 - \nu} \right) = -\zeta/Gr^3 + dN/2Gr^2$$

$$\varepsilon_\theta = \frac{1 - \nu^2}{E} \left(\sigma_\theta - \frac{\nu\sigma_r}{1 - \nu} \right) = \zeta/Gr^3 - dN/2Gr^2$$

$$\gamma_{r\theta} = \tau_{r\theta}/G = 2\eta/Gr^3$$

$$(2.29)$$

$$u_r = -\int \varepsilon_r \, dr = -\zeta/2Gr^2 + dN/2Gr + f_1(\theta)$$

$$u_\theta = -\int (u_r + r\varepsilon_\theta) \, d\theta = \eta/2Gr^2 - \int f_1(\theta) \, d\theta + f_2(r)$$

where $f_1(\theta)$ and $f_2(r)$ are arbitrary functions of θ and r. Substitution of the previous expressions into the strain displacement relation

$$\gamma_{r\theta} = \frac{u_\theta}{r} - \frac{1}{r}\frac{\partial u_r}{\partial \theta} - \frac{\partial u_\theta}{\partial r}$$

gives

$$-\frac{df_1(\theta)}{d\theta} - \int f_1(\theta)\, d\theta + f_2(r) - r\frac{df_2(r)}{dr} = 0$$

and this has the general solution

$$f_1(\theta) = k_1 \sin \theta + k_2 \cos \theta$$

$$f_2(r) = k_3 r$$

where k_1, k_2, and k_3 are constants. Converting the cylindrical components of displacement given in Equations 2.29 to rectangular components, and adding these to the values of u_x and u_y given in Equations 2.27 and 2.28, yields eventually

$$u_x = -(1/4\pi G)\{4(1 - \nu)N \ln |\sin \theta/2| + (1 - 2\nu)T[\theta - \text{sgn}(\theta)]\} + J_x$$

$$u_y = (1/4\pi G)\{(1 - 2\nu)[\theta - \text{sgn}(\theta)] - 4(1 - \nu)T \ln |\sin \theta/2|\} + J_y$$

where

$$J_x = (1 - \nu)\frac{N}{\pi G} \ln 2 - \frac{(3 - 4\nu)N \ln a}{8\pi(1 - \nu)G} - \frac{N}{16\pi(1 - \nu)G} + k_2$$

$$J_y = -(1 - \nu)\frac{T}{\pi G} \ln 2 - \frac{(3 - 4\nu)T \ln a}{8\pi(1 - \nu)G} - \frac{T}{16\pi(1 - \nu)G} + k_1$$

The terms J_x and J_y contain the arbitrary constants k_1 and k_2, and therefore are themselves arbitrary constants. They are analogous to the constant b encountered in Equation 2.5 and may be determined by making some assumption about the magnitude of the displacement at a chosen point of the domain. It is preferable, however, to accept the fact that wherever there are unbalanced line loads it is not possible to determine absolute displacements. In these circumstances, differences of displacement, or relative displacements, are all that can be calculated.

Let N and T be applied at some point C, the position of which is defined by the angle ρ, as in Figure 2.17b. If θ in the above equations is replaced by $\psi(=\theta - \rho)$, then the first equation gives the value of u_m and the second equation the value of u_n. Using vector transformation equations, expressions can now be derived for u_x and u_y in terms of F_x and F_y. By summing the results for a set of applied forces F_1, F_2, etc. one obtains the solution for a general load system. It is convenient to include here Equation 2.17 for the boundary stress reformulated in terms of F_x and F_y.

$$\sigma_\theta = \frac{1}{\pi a} \sum [F_y \cos \tfrac{1}{2}(\theta + \rho) - F_x \sin \tfrac{1}{2}(\theta + \rho)] \operatorname{cosec} \tfrac{1}{2}\psi$$

$$- \frac{3 - 4\nu}{2(1 - \nu)\pi a} \sum (F_x \cos \theta + F_y \sin \theta)$$

$$u_x = \sum (VF_y - WF_x) \tag{2.30}$$

$$u_y = -\sum (VF_x + WF_y)$$

where

$$\psi = \theta - \rho$$

$$V = \frac{1 - 2\nu}{4\pi G} [\pi \operatorname{sgn}(\psi) - \psi]$$

$$W = \frac{1 - \nu}{\pi G} \ln |\sin \tfrac{1}{2}\psi| \tag{2.31}$$

As expected, the above equations are very similar to Equation 2.13, which relates to balanced load systems. For the latter, $\sum F_x = 0$ and $\sum F_y = 0$, and then the two sets of equations are identical.

Note that for any given points specified by the angles θ and ρ, one is free to measure each angle in either the positive or negative directions, e.g. angles $2\pi/3$ and $-4\pi/3$ specify the same point. However, whatever choice is made, one must ensure that ψ lies within the range $-\pi < \psi < \pi$ so that sgn ψ has its correct value. This difficulty can be avoided by replacing sgn (ψ) by $\sin (\psi)/|\sin \psi|$, in which case there is no restriction on the choice of angles.

It is often convenient to have Equations 2.30 expressed wholly in terms of cylindrical components.

$$\sigma_\theta = \frac{1}{\pi a} \sum \left[F_t \cot \tfrac{1}{2}\psi - F_r - \frac{3 - 4\nu}{2(1 - \nu)} (F_r \cos \psi + F_t \sin \psi) \right]$$

$$u_r = \sum (HF_t - JF_r) \tag{2.32}$$

$$u_\theta = -\sum (JF_t + HF_r)$$

where

$$H = V \cos \psi - W \sin \psi$$

$$J = W \cos \psi + V \sin \psi$$

Equations 2.30 and 2.32 summarize the results of the theory for the general load case, at least as far as the boundary is concerned. Since problems of practical interest involve distributed loads rather than line loads, these equations are now reformulated accordingly.

$$\sigma_\theta = \frac{1}{\pi} \int [p_y \cos \tfrac{1}{2}(\theta + \rho) - p_x \sin \tfrac{1}{2}(\theta + \rho)] \operatorname{cosec} \tfrac{1}{2}\psi \, d\rho$$

$$- \frac{3 - 4\nu}{2(1 - \nu)\pi} \int (p_x \cos \theta + p_y \sin \theta) \, d\rho$$

$$u_x = \int (Vp_y - Wp_x)a \, d\rho \qquad (2.33)$$

$$u_y = -\int (Vp_x + Wp_y)a \, d\rho$$

$$\sigma_\theta = \frac{1}{\pi} \int \left[p_t \cot \tfrac{1}{2}\psi - p_r - \frac{3 - 4\nu}{2(1 - \nu)} (p_r \cos \psi + p_t \sin \psi) \right] d\rho$$

$$u_r = \int (Hp_t - Jp_r)a \, d\rho \qquad (2.34)$$

$$u_\theta = \int (Jp_t + Hp_r)a \, d\rho$$

where p_x, p_y, p_t, and p_r are components of the surface traction.

2.2.8 Practical examples

In illustrating the use of these equations, a useful example is that provided by the shallow tunnel. In determining the stress around a tunnel associated with the field stresses, it is usual to ignore the variation of the field stress with depth. However, when the tunnel is located at depths less than 12 times its diameter, this variation becomes important and should be allowed for in the analysis. The case which will now be considered is that of an unlined tunnel excavated in rock with a specific weight γ, there being a vertical field stress $p(= \gamma \times \text{depth})$ and a horizontal field stress Kp. The tunnel is horizontal and has its axis h metres below the ground surface, as in Figure 2.21a. Before the tunnel is excavated, the stress components at the potential tunnel boundary are

$$\sigma_r = \tfrac{1}{2}p[(1 + K) - (1 - K) \cos 2\theta]$$

$$\sigma_\theta = \tfrac{1}{2}p[(1 + K) + (1 - K) \cos 2\theta] \qquad (2.35)$$

$$\tau_{r\theta} = \tfrac{1}{2}p(1 - K) \sin 2\theta$$

where

$$p = \gamma(h - a \sin \theta)$$

As a result of excavation, the surface tractions σ_r and $\tau_{r\theta}$ are reduced to zero, and the effect of this on σ_θ can be worked out using the first of Equations 2.34. Now

$$\int (p_r \cos \psi + p_t \sin \psi) \, d\rho = \sum (F_x \cos \theta + F_y \sin \theta)/a$$

$$= \gamma \pi a \sin \theta$$

since $\sum F_x = 0$ and $\sum F_y =$ weight of rock removed per unit length of

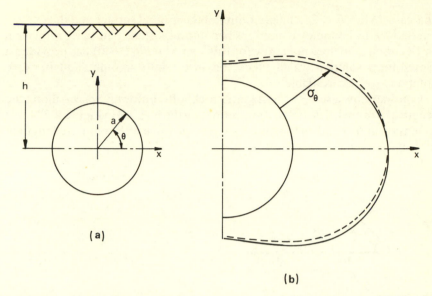

Figure 2.21   (a) A shallow tunnel with its centre at a depth h below the ground surface; (b) distribution of the boundary stress given by Equation 2.36, with and without correction term.

tunnel $= \gamma \pi a^2$. Allowing for the local normal traction effect, the stress produced by excavation is

$$\sigma'_\theta = \frac{1}{\pi} \int_{-\pi}^{\pi} \left[\tfrac{1}{2}(1 - K) \sin 2\rho \cot \tfrac{1}{2}\psi + \tfrac{1}{2}(1 + K) - \tfrac{1}{2}(1 - K) \cos 2\rho \right]$$

$$\times (h - a \sin \rho) \, d\rho - \frac{3 - 4\nu}{2(1 - \nu)} \, \gamma a \sin \theta$$

$$- \tfrac{1}{2}[(1 + K) - (1 - K) \cos 2\theta]\gamma(h - a \sin \theta)$$

Adding this change of stress to the original stress (Eqn 2.35) gives

$$\sigma_\theta = \gamma h[1 + K + 2(1 - K) \cos 2\theta] - \gamma a \sin \theta \left[\frac{3 - 4\nu}{2(1 - \nu)} + 2(1 - K) \cos 2\theta \right]$$

$$(2.36)$$

The first term is proportional to γh and is the stress which would be obtained if the gradient of the field stress with depth were ignored. The second term, prefixed by $\gamma a \sin \theta$, represents the correction which allows for the increase of field stress with depth. Equation 2.36 agrees with an equation given by Savin (1961).

Taking $h/a = 7$, $K = 0.5$, and $\nu = 0.25$, the distribution of stress around the boundary assumes the form represented by the continuous line in Figure 2.21b. The broken line shows the distribution when the correction term in Equation 2.36 is ignored. The maximum difference is of the order of 10% in this case. When $h < 7a$, the proximity of the ground surface and the control exerted by the boundary conditions become important. The correction term in Equation 2.36 does not allow for this, but Mindlin (1940) has provided a closed-form solution which takes this effect into account, and iterative solutions are also available.

Consider now a series of horizontal rockbolts uniformly spaced along the length of a tunnel. The bolts, represented by the line AB in Figure 2.22a, are in tension and are taken to impose an average force of P per metre length of

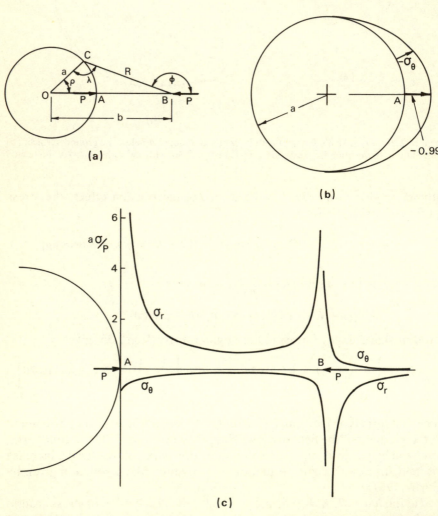

(a)

(b)

(c)

Figure 2.22 (a) Line forces P applied by a line of rockbolts, with the head at A and the anchor at B. Taking $\nu = 0.25$ and $b/a = 3$, (b) gives the distribution of the boundary stress, and (c) gives the distribution of stress along the bolt line.

the tunnel at the bolt heads and at the anchors. Treating the anchor force as a line load applied to an infinite medium, the state of stress at any boundary point C will be

$$\sigma_R = -\frac{(3 - 2\nu)P \cos \phi}{4\pi(1 - \nu)R}$$

$$\sigma_\phi = \frac{(1 - 2\nu)P \cos \phi}{4\pi(1 - \nu)R}$$

$$\tau_{R\theta} = \frac{(1 - 2\nu)P \sin \phi}{4\pi(1 - \nu)R}$$

Transforming to cylindrical components of stress with the pole at the centre O of the circle

$$\sigma'_r = -P[\cos \phi + 2(1 - \nu) \cos \phi \cos 2\lambda + (1 - 2\nu) \sin \phi \sin 2\lambda]/4\pi(1 - \nu)R$$

$$\sigma'_\theta = -P[\cos \phi - 2(1 - \nu) \cos \phi \cos 2\lambda - (1 - 2\nu) \sin \phi \sin 2\lambda]/4\pi(1 - \nu)R$$

$$\tau'_{r\theta} = P[(1 - 2\nu) \sin \phi \cos 2\lambda - 2(1 - \nu) \cos \phi \sin 2\lambda]/4\pi(1 - \nu)R$$

where $\lambda = \phi - \rho$. Equal and opposite tractions must be imposed on the boundary so as to eliminate those which have just been introduced. These produce an additional boundary stress of

$$\sigma''_\theta = -\frac{1}{\pi} \int_0^{2\pi} (\tau'_{r\theta} \cot \tfrac{1}{2}\psi - \sigma'_r) \, d\rho - \sigma'_r$$

This integral must be evaluated by numerical quadrature, as it has not been found possible to express it in closed form.

There is a further contribution to σ_θ due to the action of the bolt-head force, P,

$$\sigma'''_\theta = -\frac{P}{\pi a}\left[1 + \frac{3 - 4\nu}{2(1 - \nu)} \cos \theta\right]$$

Superposing the various contributions, the resultant boundary stress is $\sigma_\theta = \sigma'_\theta + \sigma''_\theta + \sigma'''_\theta$. With appropriate changes in the formulae, the same procedure can be used to calculate the stress components at any point in the surrounding medium. The results of such calculations are shown in Figures 2.22b and c, which show the variation of stress around the wall of the hole and along the axis of the bolt for a system having a Poisson's ratio of 0.25 and a b/a ratio of 3. It would be wrong to pay too much attention to the quantitative results of such calculations, since the averaging process used to fit the problem into the plane strain framework has the effect of smoothing out the high stress concentrations which occur in the true three-dimensional situation. In a qualitative sense, however, the model should prove useful, and the results confirm what one would expect from such a model. For instance, between A and B the radial stresses are compressive, whereas to the right of B they become tensile. The anchorage which is taken to impose a line load on the rock mass produces an infinite stress concentration. At the bolt head the radial stress is again infinite, but the circumferential stress remains finite.

2.3   Zones of influence

When an excavation is made there is a large change in the stress field in the immediate neighbourhood of the excavation. Considering points further and further from the hole, the disturbance to the stress field rapidly attenuates so that at a certain distance the effect of the excavation can be taken as negligible. The region surrounding the hole within which the disturbance is significant is termed the zone of influence. This concept is particularly useful in situations where there are two or more excavations in close proximity. Provided that none of the zones of influence overlap each other, then the stress distribution can be worked out on the basis that each of the excavations is isolated from its neighbours, the interaction between them being negligible. Furthermore, if the excavations are close enough for the zones of influence to overlap but not so close that the zones encompass any part of the boundaries of neighbouring excavations, then the resultant disturbance to the stress field can be estimated fairly accurately by superposing the disturbances produced by each of the excavations in isolation from its neighbours.

In a plane strain situation, only two principal stresses need to be considered. Let the principal field stresses be p_1 and p_3, and let the principal stresses at a typical point be σ_1 and σ_3. In considering the change of stress at the point in question, any change in the orientation of the principal stress axes should be in principle considered, but this is not so important as the change in magnitude and hence will be ignored. The changes in principal stress are $\sigma_1 - p_1$ and $\sigma_3 - p_3$ and these can be expressed as percentage changes, viz. $100(\sigma_1 - p_1)/p_1$ and $100(\sigma_3 - p_3)/p_3$. This is not satisfactory, however, as when p_3 is small (in extreme cases it might be zero!) the second of these percentages is magnified out of all proportion. It is therefore better to base both percentage changes on p_1. Furthermore, in the present context the sign of the change is not important, and hence the percentage changes can be expressed $100|\sigma_1 - p_1|/p_1$ and $100|\sigma_3 - p_3|/p_1$. The actual percentage change that one considers to be significant is a matter of personal choice. In the following discussion I have chosen a figure of 5% in fixing the boundary of the zone of influence. Thus a 5% zone of influence is defined as the region surrounding an excavation within which $100|\sigma_1 - p_1|/p_1$ or $100|\sigma_3 - p_3|/p_1$ is greater than or equal to 5%.

2.3.1 Circular hole in an isotropic medium

An example of a 5% zone of influence is shown in Figure 2.23. This is for a **circular tunnel** in an isotropic elastic rock where the original field stresses are p (vertical) and $0.5p$ (horizontal) where p is arbitrary. . Stress contours are plotted for $\sigma_1 = (1 \pm 0.05)p$ and $\sigma_3 = (0.5 \pm 0.05)p$. The zone of influence is the region contained by the outermost of these contours, and is the area shown hatched in the figure. The $0.55p$ contours are of no consequence as they lie entirely within the $0.95p$ contours. It is convenient in this and subsequent cases to omit the factor p, and to take the maximum field stress

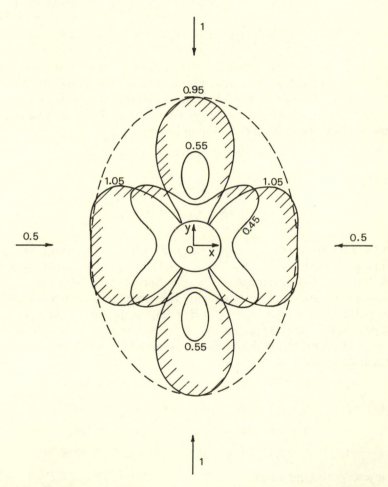

Figure 2.23 The 5% zone of influence for a circular hole in an isotropic rock where the principal field stresses are 1 (vertical) and 0.5 (horizontal). An elliptical approximation to this zone is shown by a broken line.

as unity. In many cases in practice, the precise shape of the zone of influence is not too important, and a sufficiently good approximation is to take the zone as elliptical in shape with minor and major axes coinciding with the width and height of the true zone of influence. In Figure 2.23, the approximate zone is indicated by broken lines.

Whenever possible, it is worth while adopting this type of approximation, since it avoids the tedious computational effort involved in plotting contours. Only the width and height of the zone have to be determined. In the case of the circular hole this is done very easily through the Kirsch equations:

$$\sigma_r = \tfrac{1}{2}(p_x + p_y)\left(1 - \frac{a^2}{r^2}\right) + \tfrac{1}{2}(p_x - p_y)\left(1 + 3\frac{a^4}{r^4} - 4\frac{a^2}{r^2}\right)\cos 2\theta$$

$$\sigma_\theta = \tfrac{1}{2}(p_x + p_y)\left(1 + \frac{a^2}{r^2}\right) - \tfrac{1}{2}(p_x - p_y)\left(1 + 3\frac{a^4}{r^4}\right)\cos 2\theta$$

(2.37)

where p_x, p_y are the field stresses, a = radius of the hole, r, θ = cylindrical coordinates. In locating the points where the zone boundary cuts the x axis ($\theta = 0°$), it is necessary to consider four possibilities: $\sigma_r = p_x + 0.05p$ and $\sigma_\theta = p_y \pm 0.05p$, where p is the larger of p_x and p_y. Each of these requirements produces a quadratic equation in (a^2/r^2):

$$A(a/r)^4 + B(a/r)^2 + C = 0$$

with roots

$$\frac{a}{r} = \left\{\frac{1}{2A}\left[B \pm (B^2 - 4AC)^{1/2}\right]\right\}^{1/2}$$

In the first two cases $A = 3(p_x - p_y)$, $B = 3p_y - 5p_x$, and $C = \pm p/10$, and in the second two cases $A = 3(p_y - p_x)$, $B = p_x + p_y$, and $C = \pm p/10$. Consideration of the intercepts of the zone boundary with the y axis leads to a quadratic equation of the same form. This time there are two possibilities corresponding to $A = 3(p_y - p_x)$, $B = 3p_x - 5p_y$, and $C = \pm p/10$, and two possibilities corresponding to $A = 3(p_x - p_y)$, $B = p_x + p_y$, and $C = \pm p/10$. Considering the previous example, $p_x = 0.5$ and $p_y = 1$, and for the intercepts on the x axis:

(a) $A = -1.5$ $B = 0.5$ $C = 0.1$ real root, $r = 1.45a$

(b) $A = -1.5$ $B = 0.5$ $C = -0.1$ complex roots only

(c) $A = 1.5$ $B = 1.5$ $C = 0.1$ complex roots only

(d) $A = 1.5$ $B = 1.5$ $C = -0.1$ real root, $r = 3.99a$

For intercepts on the y axis:

(a) $A = 1.5$ $B = -3.5$ $C = 0.1$ real roots, $r = 5.88a, 0.659a$

(b) $A = 1.5$ $B = -3.5$ $C = -0.1$ real root, $r = 0.651a$

(c) $A = -1.5$ $B = $ 1.5 $C = $ 0.1   real root, $r = 3.73a$

(d) $A = -1.5$ $B = $ 1.5 $C = -0.1$ real root, $r = 0.97a$

The major and minor axes are therefore $11.76a$ and $7.98a$ respectively.

2.3.2   Elliptical hole in an isotropic medium

Consideration will now be given to an **elliptical hole** in an isotropic rock under conditions of plane strain. Unfortunately most textbooks which deal with the stress distribution around an elliptical hole restrict attention to the stress at the boundary. General expressions for the stress components at an arbitrary point can be developed in the following manner.

Figure 2.24a shows an elliptical hole with semi-axes a and b in a uniaxial stress field p which is inclined at angle β to the horizontal. An elliptical coordinate system is established with coordinate lines consisting of confocal ellipses $\xi = $ constant and confocal hyperbolas $\eta = $ constant (Fig. 2.24b). The transformation equations relating elliptic coordinates to rectangular coordinates are

$$x = A \cos \eta, \qquad y = B \sin \eta \tag{2.38}$$

where

$$A = c \cosh \xi, \qquad B = c \sinh \xi \tag{2.39}$$

and

$$c^2 = A^2 - B^2 = a^2 - b^2 \tag{2.40}$$

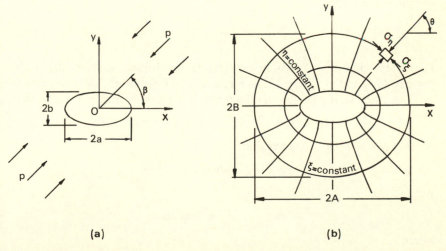

(a)   (b)

Figure 2.24   (a) An elliptical excavation in a uniaxial stress field p; (b) an elliptical coordinate system confocal with the elliptical hole.

A and B are the semi-axes of the ellipse which passes through point (x, y). The particular ellipse which coincides with the boundary of the hole is taken to be $\xi = \xi_0$. The particular ellipse which passes through a given point (x, y) is found as follows. From Equations 2.38,

$$x^2/A^2 + y^2/B^2 = 1$$

Replacing A^2 by $a^2 - b^2 + B^2$, and rearranging gives

$$B^4 - 2fB^2 - t = 0$$

where

$$f = \tfrac{1}{2}(x^2 + y^2 + b^2 - a^2) \tag{2.41}$$

and

$$t = (a^2 - b^2)y^2 \tag{2.42}$$

Hence

$$B = [f + \sqrt{(f^2 + t)}]^{1/2} \tag{2.43}$$

and

$$A = (a^2 - b^2 + B^2)^{1/2} \tag{2.44}$$

The direction of the tangent to the ellipse passing through the general point (x, y) is given by

$$\frac{dy}{dx} = -\frac{B}{A} \cot \eta = -\frac{BX}{AY}$$

where $X = x/A$ and $Y = y/B$. The angle between the normal to the ellipse and the x axis is then

$$\alpha = \arctan \frac{AY}{BX} \tag{2.45}$$

The stress analysis equations which are to follow involve the terms $e = \exp(2\xi)$ and $e_0 = \exp(2\xi_0)$. Referring to Equations 2.39 and expressing $\cosh \xi$ and $\sinh \xi$ in exponential form, it is a simple matter to show that

$$e = (A + B)/(A - B), \qquad e_0 = (a + b)/(a - b) \tag{2.46}$$

Writing $\zeta = \xi + i\eta$, it follows that

$$\cosh \zeta = \lambda \cosh \eta + i\mu \sin \eta$$

and   (2.47)

$$\sinh \zeta = \mu \cos \eta + i\lambda \sin \eta$$

where

$$\lambda = \tfrac{1}{2}(e + 1)/\sqrt{e}, \qquad \mu = \tfrac{1}{2}(e - 1)/\sqrt{e}$$

The normal procedure for solving stress analysis problems in curvilinear coordinates is through the Kolosoff equations:

$$\sigma_\xi + \sigma_\eta = 2[\psi'(z) + \overline{\psi}'(\bar{z})]$$

$$\sigma_\eta - \sigma_\xi + 2i\tau_{\xi\eta} = 2e^{2i\alpha}[\bar{z}\psi''(z) + \chi''(z)] \tag{2.48}$$

$$2G(u_\xi - iu_\eta) = e^{i\alpha}[(3 - 4\nu)\overline{\psi}(\bar{z}) - \bar{z}\psi'(z) - \chi'(z)]$$

In these expressions $\psi(z)$ and $\chi(z)$ are complex stress functions which have to be chosen to satisfy the boundary conditions, and α is the inclination of the $\eta = $ constant coordinate line to the x axis.

In the present case, it is found that (Timoshenko & Goodier 1951)

$$\psi(z) = \tfrac{1}{4}pc\{\exp (2\xi_0) \cos 2\beta \cosh \zeta + [1 - \exp (2\xi_0 + 2i\beta)] \sinh \zeta\}$$

$$\chi(z) = -\tfrac{1}{4}pc^2[(\cosh 2\xi_0 - \cos 2\beta)\zeta + \tfrac{1}{2} \exp (2\xi_0) \cosh 2(\zeta - \xi_0 - i\beta)] \tag{2.49}$$

The stress and displacement at any point (x, y) can be determined by substituting the above expressions for $\psi(z)$ and $\chi(z)$ into Equations 2.48. This process is lengthy and tedious but only involves simple complex algebra in sorting out the real and imaginary parts with the assistance of Equations 2.47. Restricting attention to stresses, the results are as follows:

$$\frac{\sigma_\xi}{p} = \frac{e - e_0}{J^2} \left\{ (ee_0 - 1)\left(\cos 2\beta + \frac{e^2 - 1}{2e_0} \right) \right.$$

$$\left. + \left[\frac{J}{2}(e - e_0) + e(1 - ee_0) \right] \cos 2(\beta - \eta) \right\}$$

$$\frac{\sigma_\eta}{p} = \frac{1}{J}[2e_0 \cos 2\beta + (e^2 - 1) - 2ee_0 \cos 2(\beta - \eta)] - \sigma_\xi \tag{2.50}$$

$$\frac{\tau_{\xi\eta}}{p} = \frac{e - e_0}{J^2} \left\{ \left[\frac{J}{2}(e_0 + e) + e^2 e_0 \right] \sin 2(\beta - \eta) + e \sin 2(\beta + \eta) \right.$$

$$\left. + \frac{e}{e_0}(ee_0 - 1) \sin 2\eta - e(e + e_0) \sin 2\beta \right\}$$

where

$$J = e^2 + 1 - 2e \cos 2\eta$$

A more general case is shown in Figure 2.25 where there are field stresses p_x, p_y and p_{xy}. In this situation the stress distribution can be determined from Equations 2.50 by superposition of the following four cases:

(a) $p = p_x,$ $\beta = 0$
(b) $p = p_y,$ $\beta = \pi/2$
(c) $p = p_{xy},$ $\beta = \pi/4$
(d) $p = -p_{xy},$ $\beta = -\pi/4$

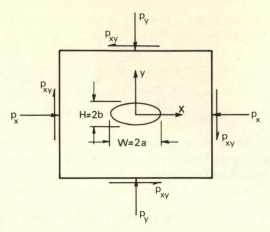

Figure 2.25 An elliptical opening in a general stress field subject to the conditions of plane strain.

One further stage of generalization can be undertaken by considering, in addition to the above, the antiplane field stresses p_{yz} and p_{zx}. When these are included, as in Figure 2.26, we have the most general case of axially homogeneous strain, the state of stress and strain varying only across the transverse section of the excavation and not along its longitudinal axis. It is not possible at this stage to go into all the details of the analysis, but the final results are now presented, both for stress and displacement. The complete set of equations are given so that the reader can encode it for use with a computer, should that be required.

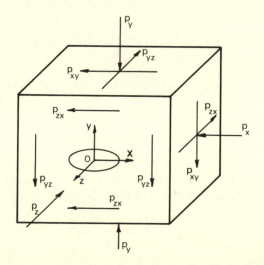

Figure 2.26 An elliptical opening in a general stress field subject to the conditions of axially homogeneous strain.

$$t = (a^2 - b^2)y^2, \qquad f = \tfrac{1}{2}(x^2 + y^2 + b^2 - a^2)$$

$$B = [f + (f^2 + e)^{1/2}]^{1/2}, \qquad A = (a^2 - b^2 + B^2)^{1/2}$$

$$X = x/A, \qquad Y = y/B$$

$$\alpha = \arctan \frac{AY}{BX}, \qquad S = 2XY, \qquad C = 1 - 2Y^2$$

$$d = a + b, \qquad D = A + B, \qquad h = A - B$$

$$j = d^2 - h^2, \qquad k = d^2 + h^2$$

$$q = (D^2 + d^2)/D, \qquad g = (D^2 - d^2)/D$$

$$\Delta = 2(B^2X^2 + A^2Y^2), \qquad L = 2gD/\Delta$$

$$J = 2(AB - ab)d/D\Delta, \qquad T = h(g + D)/d$$

$$M = (3 - 4\nu), \qquad N = ghD/\Delta d$$

$$P = bp_x + ap_y, \qquad Q = bp_x - ap_y$$

$$\sigma_\xi = \frac{g}{4\Delta^2} \left\{ \frac{2ABDj}{d^2}(p_x + p_y) + [(jD - \Delta g)C - jh] \right.$$
$$\left. (p_y - p_x) - 2(jD - \Delta g)Sp_{xy} \right\}$$

$$\sigma_\eta = \frac{2}{\Delta} \left\{ AB(p_x + p_y) + \frac{1}{2}\left(C - \frac{h}{D}\right)d^2(p_y - p_x) - d^2 Sp_{xy} \right\} - \sigma_\xi$$

$$\sigma_z = p_z + \nu(\sigma_\xi + \sigma_n - p_x - p_y)$$

$$\tau_{\xi\eta} = \frac{g}{4\Delta^2} \left\{ \frac{jhD^2S}{d^2}(p_x + p_y) + (jD + \Delta q)S(p_y - p_x) \right.$$
$$\left. + 2[(kD + \Delta q)C - qDh]p_{xy} \right\}$$

$$\tau_{z\xi} = \frac{g}{\sqrt{2}\Delta}(Yp_{yz} + Xp_{zx})$$

$$\tau_{\eta z} = \frac{q}{\sqrt{2}\Delta}(Xp_{yz} - Yp_{zx})$$

$$u_x = \frac{-d}{4GD} \left[\left\{ P + Q\left[M + \left(\frac{Bb}{Dd} - Y^2\right)L \right]\right\}X \right.$$
$$\left. + [Md + T - NA + (B + DC)J]Yp_{xy} \right]$$

$$u_y = \frac{-d}{4GD} \left[\left\{ P - Q\left[M + \left(\frac{Aa}{Dd} - X^2\right)L \right]\right\}Y \right.$$
$$\left. + [Md - T + NB + (A - DC)J]Xp_{xy} \right]$$

$$u_z = \frac{-d}{GD}(bXp_{zx} + aYp_{yz})$$

It is to be noted that whereas the given stress components are total stress components (i.e. those existing after excavation), the given displacement components are those induced by the excavation and not those which would result if the rock were loaded from a state of zero stress.

Adopting a trial and error procedure, the foregoing equations may be used to search for and establish points on the boundary of the zone of influence. The shape and size of the zone will depend on the magnitude and orientation of the principal field stresses (p_1, p_3) and on the width:height ratio (q) of the excavation. Although the above analysis allows for antiplane components of field stress, the present investigation has been limited to plane strain conditions. Plots of the stress contours corresponding to a 5% deviation from the principal field stresses are shown in Figures 2.27, 2.28, and 2.29 for three typical cases. The data for these figures are as follows:

Figure	Width:height ratio q	Field stresses p_1	p_3	Angle between p_1 and x axis
2.27	2	1	0.1	$\pi/2$
2.28	2	1	0.5	$\pi/2$
2.29	2	1	0.5	$\pi/4$

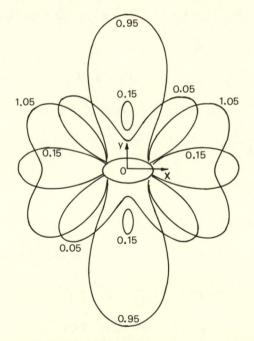

Figure 2.27 Stress contours for an elliptical hole with a width:height ratio of 2, when the principal field stresses are 1 vertical and 0.1 horizontal.

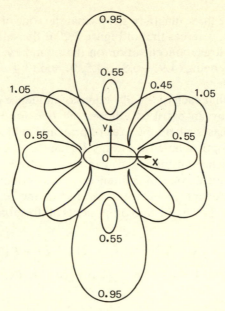

Figure 2.28 Stress contours for an elliptical hole with a width : height ratio of 2, when the principal field stresses are 1 vertical and 0.5 horizontal.

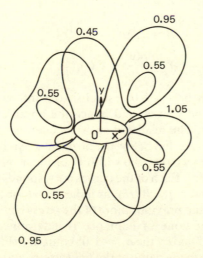

Figure 2.29 Stress contours for an elliptical hole with a width : height ratio of 2, when the principal field stresses are 1 and 0.5, inclined ±45° to the horizontal.

Comparison of the three diagrams shows that the zone of influence of Figure 2.27 is the largest, whereas that of Figure 2.29 is the smallest. This accords with the level of stress concentration on the boundary, the maximum and minimum stresses being (4.9, −0.8), (4.5, 0) and (3.47, 0.28) for the three cases taken in order.

It will be seen that each of the 5% zones of influence can be represented reasonably well by an escribed ellipse, the one for Figure 2.29 having inclined axes. In cases where the axes of principal field stress coincide with the axes of the elliptical hole (as in Figs 2.27 & 2.28), the intercepts of the true zone of influence on these axes can be used to define the axes of the approximate elliptical zone of influence, as was done previously for the circular hole. Referring to Figure 2.25, and taking $p_{xy} = 0$, $p_y = p$, and $p_x = Kp$, it can be shown that the components of stress on the horizontal and vertical axes are

$$\sigma_x = \{K - [q^2 + (1 - 2q)K]A - (q - K)B\}p$$
$$\sigma_y = \{1 - (q^2 + K - 2q)A + (q - K)B\}p \tag{2.51}$$

where on the x axis

$$A = \left(1 - \frac{2x}{HX}\right) \Big/ (q - 1)^2$$

$$B = 2x/(q - 1)HX^3 \tag{2.52}$$

$$X = \left(\frac{4x^2}{H^2} - q^2 + 1\right)^{1/2}$$

and on the y axis

$$A = \left(1 - \frac{2y}{HY}\right) \Big/ (q - 1)^2$$

$$B = 2q^2y/(1 - q)HY^3 \tag{2.53}$$

$$Y = \left(\frac{4y^2}{H^2} - 1 + q^2\right)^{1/2}$$

These equations can be used to establish the intercepts of the zone of influence on the x and y axes, and hence the approximate width W_i and height H_i of the zone. The results of such calculations are plotted in Figure 2.30, the ordinates above and below the horizontal axis being W_i/L and H_i/L respectively, where L is either the width W or the height H of the opening, whichever is the greater. Each of the curves corresponds to a particular value of the width: height ratio $q(=W/H)$, as indicated at the sides of the figure.

Referring back to the previous plots of the stress contours, it will be seen that the width of the zone of influence is determined by the $\sigma_3 = 0.15$ contour in Figure 2.27 and by the $\sigma_1 = 1.05$ contour in Fig. 2.28. This change results from a change in the value of the field stress ratio K, and it is this kind of occurrence which produces the kinks in the curves of Figure 2.30.

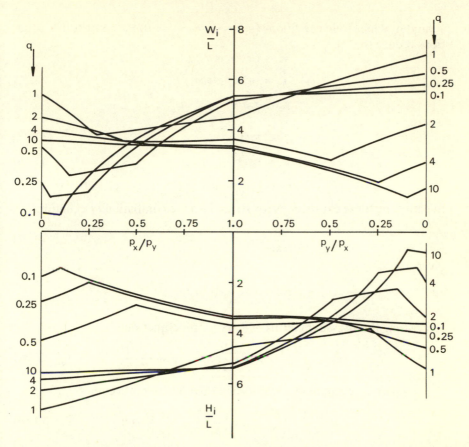

Figure 2.30 Diagram showing how the width W_i and height H_i of the zone of influence of an elliptical excavation in an isotropic rock varies with the width : height ratio of the excavation and the field stress ratio $p_x : p_y$ (or $p_y : p_x$).

2.3.3 Elliptical hole in a transversely isotropic medium

Many rocks have a laminar structure and consequently exhibit transverse isotropy in their elastic properties. An axis perpendicular to the planes of stratification is an axis of symmetry. If y is an axis of symmetry (Fig. 2.31) then the constitutive equations may be written

$$\varepsilon_x = (\sigma_x - \nu_0\sigma_y - \nu\sigma_z)/E: \qquad \gamma_{xy} = \tau_{xy}/N$$
$$\varepsilon_y = (n\sigma_y - \nu_0\sigma_x - \nu_0\sigma_z)/E: \qquad \gamma_{yz} = \tau_{yz}/N \qquad (2.54)$$
$$\varepsilon_z = (\sigma_z - \nu\sigma_x - \nu_0\sigma_y)/E: \qquad \gamma_{xz} = \tau_{zx}/G$$

where E and E/n are elastic moduli, G and N are shear moduli, ν and ν_0 are Poisson's ratios, and $G = E/2(1 + \nu)$.

Under plane strain conditions ($\varepsilon_z = \gamma_{zx} = \gamma_{yz} = 0$) the constitutive equations reduce to

$$\varepsilon_x = (\sigma_x - \nu_0'\sigma_y)/E'$$
$$\varepsilon_y = (n'\sigma_y - \nu_0'\sigma_x)/E' \qquad (2.55)$$
$$\gamma_{xy} = \tau_{xy}/N$$

where

$$E' = E/(1 - \nu^2)$$
$$\nu_0' = \nu_0/(1 - \nu) \qquad (2.56)$$
$$n' = (n - \nu_0^2)/(1 - \nu^2)$$

Substituting these expressions for strain into the compatibility equation

$$\frac{\partial^2\varepsilon_x}{\partial y^2} + \frac{\partial^2\varepsilon_y}{\partial x^2} = \frac{\partial^2\gamma_{xy}}{\partial x\,\partial y} \qquad (2.57)$$

gives

$$\frac{1}{E'}\frac{\partial^2}{\partial y^2}(\sigma_x - \nu_0'\sigma_y) + \frac{1}{E'}\frac{\partial^2}{\partial x^2}(n'\sigma_y - \nu_0'\sigma_x) = \frac{1}{N}\frac{\partial^2\tau_{xy}}{\partial x\,\partial y}$$

Expressing the stresses in terms of Airy's stress function U

$$\sigma_x = \frac{\partial^2 U}{\partial y^2}, \qquad \sigma_y = \frac{\partial^2 U}{\partial x^2}, \qquad \tau_{xy} = -\frac{\partial^2 U}{\partial x\,\partial y} \qquad (2.58)$$

the compatibility equation, after some rearrangement, becomes

$$n'\frac{\partial^4 U}{\partial x^4} + \left(\frac{E'}{N} - 2\nu_0\right)\frac{\partial^4 U}{\partial x^2\,\partial y^2} + \frac{\partial^4 U}{\partial y^4} = 0 \qquad (2.59)$$

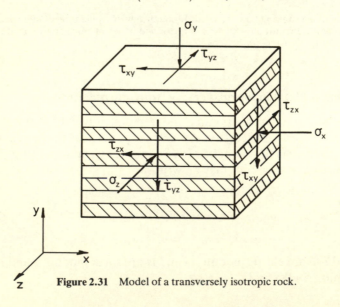

Figure 2.31   Model of a transversely isotropic rock.

This may be written

$$\left(\beta_1^2 \frac{\partial^2}{\partial x^2} + \frac{\partial^2}{\partial y^2}\right)\left(\beta_2^2 \frac{\partial^2}{\partial x^2} + \frac{\partial^2}{\partial y^2}\right) U = 0$$

where

$$\beta_1^2 \beta_2^2 = n'$$

$$\beta_1^2 + \beta_2^2 = \frac{E'}{N} - 2\nu_0 \qquad (2.60)$$

It follows that the general solution may be written

$$U = 2 \, \text{Re} \, [F_1(z_1) + F_2(z_2)] \qquad (2.61)$$

where

$$z_1 = x + iy_1 = x + i\beta_1 y$$
$$z_2 = x + iy_2 = x + i\beta_2 y$$

(2.62)

and $F_1(z_1)$, $F_2(z_2)$ are analytical functions of z_1, z_2. Let

$$\frac{dF_1}{dz_1} = \phi(z_1) \quad \text{and} \quad \frac{dF_2}{dz_2} = \psi(z_2)$$

Then

$$\frac{\partial^2 F_1}{\partial x^2} = \left(\frac{\partial z_1}{\partial x}\right)^2 \phi'(z_1) = \phi'(z_1)$$

$$\frac{\partial^2 F_1}{\partial y^2} = \left(\frac{\partial z_1}{\partial y}\right)^2 \phi'(z_1) = -\beta_1^2 \phi'(z_1)$$

$$\frac{\partial^2 F_1}{\partial x \, \partial y} = \left(\frac{\partial z_1}{\partial x} \frac{\partial z_1}{\partial y}\right) \phi'(z_1) = i\beta_1 \phi'(z_1)$$

There are similar expressions for the partial derivatives of ψ. Using these and substituting for U from Equation 2.61 into Equations 2.58, produces Lekhnitskii's equations

$$\sigma_x = -2 \, \text{Re} \, [\beta_1^2 \phi'(z_1) + \beta_2^2 \psi'(z_2)]$$

$$\sigma_y = 2 \, \text{Re} \, [\phi'(z_1) + \psi'(z_2)] \qquad (2.63)$$

$$\tau_{xy} = 2 \, \text{Im} \, [\beta_1 \phi'(z_1) + \beta_2 \psi'(z_2)]$$

The complex functions $\phi(z)$ and $\psi(z)$ must be chosen to satisfy the boundary conditions, and once this is done, Equations 2.63 can be used to determine the state of stress at all points.

An important point revealed by Equations 2.63 is that the stress distribution is dependent on the two elastic constants β_1 and β_2, these being derived from the usual constants through Equations 2.56 and 2.60. This compares with the homogeneous isotropic case where the stress is independent of elastic constants.

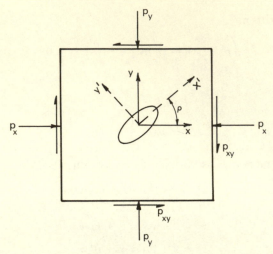

Figure 2.32   An inclined elliptical excavation in a general stress field subject to the conditions of plane strain.

The method will now be applied to finding the stress distribution around an elliptical excavation. As in the isotropic case, the literature in general avoids the calculation of stresses within the rock mass and focuses attention on the boundary stresses only. Since the present interest lies in the extent to which the disturbance produced by an excavation penetrates the country rock, it is necessary to develop the relevant equations. The procedure adopted is virtually the same as that used by Savin (1961), and to assist in correlation the same notation will be employed here.

The elliptical opening shown in Figure 2.32 has principal axes $2a$ and $2b$, the $2a$ axis being inclined at angle ρ to the x axis. The y axis is the axis of symmetry of the rock structure. The field stress components are p_x, p_y, and p_{xy}. Before proceeding with the stress analysis, there are a number of points relating to the geometry of the ellipse which have to be established. With coordinate axes x' and y' aligned parallel to the principal axes of the ellipse, the ellipse can be represented in parametric form as follows:

$$x' = a \cos \theta = \tfrac{1}{2}a(\sigma + \sigma^{-1})$$
$$y' = -b \sin \theta = \tfrac{1}{2}ib(\sigma - \sigma^{-1})$$

where

$$\sigma = e^{i\theta}$$

(Note that if the real and imaginary parts of σ are treated as coordinates, then they trace out a circle of unit radius.)

The horizontal and vertical coordinates of a point on the ellipse are

$$x = x' \cos \rho - y' \sin \rho, \qquad y = y' \cos \rho + x' \sin \rho$$

or

$$x = \tfrac{1}{2}[(a \cos \rho - ib \sin \rho)\sigma + (a \cos \rho + ib \sin \rho)\sigma^{-1}]$$

$$y = \tfrac{1}{2}[(a \sin \rho + ib \cos \rho)\sigma + (a \sin \rho - ib \cos \rho)\sigma^{-1}]$$

(2.64)

Then

$$z = x + iy = \tfrac{1}{2}[(a - b)\sigma + (a + b)\sigma^{-1}]e^{i\rho}$$

Consider the function

$$z = \omega(\zeta) = \tfrac{1}{2}[(a - b)\zeta + (a + b)\zeta^{-1}]e^{i\rho}$$

where

$$\zeta = \xi + i\eta$$

This function maps the region outside the ellipse in the x, y plane (Fig. 2.33a) on to the inside of the unit circle on the ξ, η plane (Fig. 2.33b). Using Equations 2.64 it can be shown that

$$z_1 = \tfrac{1}{2}[(A_1 + iS_1)\sigma + (A_1 - iS_1)\sigma^{-1}]$$

$$z_2 = \tfrac{1}{2}[(A_2 + iS_2)\sigma + (A_2 - iS_2)\sigma^{-1}]$$

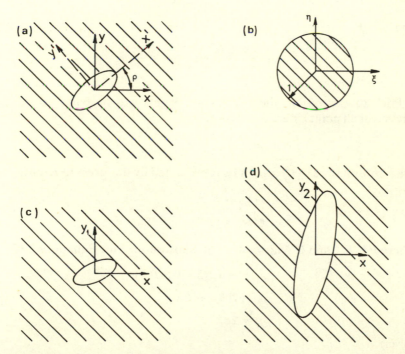

Figure 2.33 (a) An inclined elliptical excavation in a transversely isotropic rock; (b) the unit circle in the $\xi\eta$ plane produced by applying the mapping function $z = w(\zeta)$ to the ellipse in (a); (c) the ellipse in the x,y_1 plane produced by applying the mapping function $z = x + i\beta_{1y}$ to the ellipse in (a); (d) the ellipse in the x,y_2 plane produced by applying the mapping function $z = x + i\beta_{2y}$ to the ellipse in (a).

where

$$A_1 = a(\cos \rho + i\beta_1 \sin \rho), \qquad S_1 = b(i\beta_1 \cos \rho - \sin \rho)$$
$$A_2 = a(\cos \rho + i\beta_2 \sin \rho), \qquad S_2 = b(i\beta_2 \cos \rho - \sin \rho)$$

$$(2.65)$$

In the (x, y_1) coordinate system the elliptical hole plots as another ellipse with all the vertical ordinates multiplied by β_1 (Fig. 2.33c). The function

$$z_1 = \omega_1(\zeta) = \tfrac{1}{2}[(A_1 + iS_1)\zeta + (A_1 - iS_1)\zeta^{-1}] \qquad (2.66)$$

maps the area outside this new ellipse on to the inside of the unit circle. A similar interpretation applies to the function

$$z_2 = \omega_2(\zeta) = \tfrac{1}{2}[(A_2 + iS_2)\zeta + (A_2 - iS_2)\zeta^{-1}] \qquad (2.67)$$

The inverses of Equations 2.66 and 2.67 are obtained by multiplying throughout by ζ and then solving the quadratic equations produced. This gives

$$\zeta = \frac{z \pm (z^2 - A_1^2 - S_1^2)^{1/2}}{A_1 + iS_1}$$

and $$(2.68)$$

$$\zeta = \frac{z \pm (z^2 - A_2^2 - S_2^2)^{1/2}}{A_2 + iS_2}$$

Prior to excavation, the stress components will be equal to the field stresses at all points, i.e.

$$\sigma_x = p_x, \qquad \sigma_y = p_y, \qquad \tau_{xy} = p_{xy}$$

This uniform state of stress can be represented by the stress functions

$$\phi(z_1) = B^* z_1$$
$$\psi(z_2) = (B'^* + iC'^*) z_2$$

$$(2.69)$$

where B^*, B'^* and C'^* are constants. Substituting into Equations 2.63,

$$p_x = -2(\beta_1^2 B^* + \beta_2^2 B'^*)$$
$$p_y = 2(B^* + B'^*)$$
$$p_{xy} = 2\beta_2 C'^*$$

Solving,

$$B^* = -\tfrac{1}{2}(p_x + \beta_2^2 p_y)/(\beta_1^2 - \beta_2^2)$$
$$B'^* = \tfrac{1}{2}(p_x + \beta_1^2 p_y)/(\beta_1^2 - \beta_2^2) \qquad (2.70)$$
$$C'^* = \tfrac{1}{2} p_{xy}/\beta_2$$

Consider any part PQ of the potential hole boundary before it is mined. Let X, Y be the components of force on PQ provided by the internal rock mass. Then, ignoring constants which have no effect on the final result,

$$X = 2 \text{ Re } [i\beta_1\phi(z_1) + i\beta_2\psi(z_2)]$$
$$Y = -2 \text{ Re } [\phi(z_1) + \psi(z_2)]$$

(2.71)

Replacing $\phi(z_1)$ and $\psi(z_2)$ by the expressions in Equations 2.69, and then using in turn Equations 2.66 and 2.67 for z_1 and z_2, Equation 2.65 for A_1, S_1, A_2, and S_2, and Equations 2.70 for B^*, B'^*, and C'^*, the expressions for X and Y eventually reduce to

$$X = K_5\sigma + \bar{K}_5\sigma^{-1}$$
$$Y = K_6\sigma + \bar{K}_6\sigma^{-1}$$

(2.72)

where

$$K_5 = \tfrac{1}{2}p_x(a \sin \rho + ib \cos \rho) - \tfrac{1}{2}p_{xy}(a \cos \rho - ib \sin \rho)$$
$$K_6 = -\tfrac{1}{2}p_y(a \cos \rho - ib \sin \rho) + \tfrac{1}{2}p_{xy}(a \sin \rho + ib \cos \rho)$$

(2.73)

and the bar denotes the complex conjugate.

After the hole is excavated, additional terms $\phi_0(z_1)$ and $\psi_0(z_2)$ must be added to the stress functions to allow for the stresses induced by the excavation. Then

$$\phi(z_1) = B^*z_1 + \Phi_0(z_1)$$
$$\psi(z_2) = (B'^* + iC'^*)z_2 + \Psi_0(z_2)$$

The additional functions must be analytical up to infinity, and can be represented by the series

$$\phi_0(z_1) = a_0 + a_{-1}z_1^{-1} + a_{-2}z_1^{-2} + \cdots$$
$$\psi_0(z_2) = b_0 - a_{-1}z_2^{-1} + a_{-2}z_2^{-2} + \cdots$$

When these are used in place of $\phi(z_1)$ and $\psi(z_2)$ in Equations 2.71, one gets the boundary force components (X_0, Y_0) generated by the excavation. Using Equations 2.66 and 2.67 to express z_1 and z_2 in terms of ζ, Equations 2.71 may be written

$$X_0 = 2 \text{ Re } [i\beta_1\Phi_0(\zeta) + i\beta_2\Psi_0(\zeta)]$$
$$Y_0 = -2 \text{ Re } [\Phi_0(\zeta) + \Psi_0(\zeta)]$$

(2.74)

where

$$\Phi_0(\zeta) = \phi_0[\omega_1(\zeta)]$$
$$\Psi_0(\zeta) = \psi_0[\omega_2(\zeta)]$$

As the surface traction on the boundary is zero, the additional boundary forces must be equal and opposite to those given in Equation 2.72, and so

$$X_0 = -(K_5\sigma + \bar{K}_5\sigma^{-1})$$
$$Y_0 = -(K_6\sigma + \bar{K}_6\sigma^{-1})$$

(2.75)

To proceed with the analysis requires the application of the Schwarz formula (Milne-Thomson 1938)

$$F(\zeta) = \frac{1}{2\pi} \int_0^{2\pi} u(\theta) \frac{\sigma + \zeta}{\sigma - \zeta} d\theta$$

(2.76)

where $F(\zeta)$ is a function which is analytic inside the unit circle, and $u(\theta)$ is the real part of $F(\zeta)$ on the circumference of the circle. Applying this formula to the first of Equations 2.74, and using 2.75,

$$i\beta_1\Phi_0(\zeta) + i\beta_2\Psi_0(\zeta) = \frac{1}{4\pi} \int_0^{2\pi} X_0 \frac{\sigma + \zeta}{\sigma - \zeta} d\theta$$

$$= -\frac{1}{4\pi} \int_0^{2\pi} (K_5\sigma + \bar{K}_5\sigma^{-1}) \frac{\sigma + \zeta}{\sigma - \zeta} d\theta$$

Now

$$\int_0^{2\pi} \frac{\sigma + \zeta}{\sigma - \zeta} \sigma \, d\theta = 4\pi\zeta$$

and

$$\int_0^{2\pi} \frac{\sigma + \zeta}{\sigma - \zeta} \frac{1}{\sigma} d\theta = 0$$

Therefore

$$\beta_1\Phi_0(\zeta) + \beta_2\Psi_0(\zeta) = iK_5\zeta$$

(2.77)

Treating the second of Equations 2.74 in exactly the same way:

$$\Phi_0(\zeta) + \Psi_0(\zeta) = K_6\zeta$$

(2.78)

Solving the simultaneous Equations 2.77 and 2.78:

$$\Phi_0(\zeta) = (iK_5 - \beta_2K_6)\zeta/(\beta_1 - \beta_2)$$

$$\Psi_0(\zeta) = (iK_5 - \beta_1K_6)\zeta/(\beta_2 - \beta_1)$$

Substituting expressions for K_5, K_6, and ζ, as given by Equations 2.73 and 2.68, the stress functions are reduced to their final form, expressed in terms of z_1 and z_2:

$$\phi_0(z_1) = \frac{\{(a \cos \rho - ib \sin \rho)(\beta_2 p_y - ip_{xy}) - (a \sin \rho + ib \cos \rho)(\beta_2 p_{xy} - ip_x)\}}{2(A_1 + iS_1)(\beta_1 - \beta_2)}$$

$$\times [z_1 - (z_1^2 - A_1^2 - S_1^2)^{1/2}]$$

$$\psi_0(z_2) = \frac{\{(a \sin \rho + ib \cos \rho)(\beta_1 p_{xy} - ip_x) - (a \cos \rho - ib \sin \rho)(\beta_1 p_y - ip_{xy})\}}{2(A_2 + iS_2)(\beta_1 - \beta_2)}$$

$$\times [z_2 - (z_2^2 - A_2^2 - S_2^2)^{1/2}]$$

(2.79)

The stress components induced by the excavation can now be obtained through the use of Equations 2.63. This requires the derivatives of $\phi_0(z_1)$ and $\psi_0(z_2)$ rather than the functions themselves. This merely amounts to replacing $[z_1 - (z_1^2 - A_1^2 - S_1^2)^{1/2}]$ and $[z_2 - (z_2^2 - A_2^2 - S_2^2)^{1/2}]$ in Equations 2.79 by $[1 - z_1(z_1^2 - A_1^2 - S_1^2)^{-1/2}]$ and $[1 - z_2(z_2^2 - A_1^2 - S_2^2)^{-1/2}]$ respectively. The resultant state of stress is obtained by adding the induced stresses to the field stresses.

For the purpose of computation, the various factors of $\Phi_0'(z_1)$ and $\psi_0'(z_2)$ have to be resolved into their real and imaginary parts. Thus, considering $\phi_0'(z_1)$,

$$(A_1 + iS_1)^{-1} = (J_1 + iJ_2)/(J_1^2 + J_2^2)$$

where

$$J_1 = (a - \beta_1 b) \cos \rho$$

$$J_2 = (b - \beta_1 a) \sin \rho$$

The terms in braces $= iJ_3 - J_4$, where

$$J_3 = (ap_x - \beta_2 bp_y) \sin \rho - (a + \beta_2 b)p_{xy} \cos \rho$$

$$J_4 = (bp_x - \beta_2 ap_y) \cos \rho + (b + \beta_2 a)p_{xy} \sin \rho$$

The square root term is

$$\sqrt{(J_7 + iJ_8)} = J_9 + iJ_{10}$$

where

$$J_7 = x^2 - \beta_1^2 y^2 + (\beta_1^2 b^2 - a^2) \cos^2 \rho + (\beta_1^2 a^2 - b^2) \sin^2 \rho$$

$$J_8 = 2\beta_1[xy + (b^2 - a^2) \sin \rho \cos \rho]$$

$$J_9 = r \cos \varepsilon$$

$$J_{10} = r \sin \varepsilon$$

$$\varepsilon = \tfrac{1}{2} \text{ATAN2} \, (J_8/J_7)$$

$$r = \pm(J_7^2 + J_8^2)^{1/4}$$

Finally

$$\phi_0'(z_1) = J_5 J_{13} - J_6 J_{14}$$

where

$$J_5 = \frac{(J_1 J_4 + J_2 J_3)}{(\beta_2 - \beta_1)(J_1^2 + J_2^2)}$$

$$J_6 = \frac{(J_2 J_4 - J_1 J_3)}{(\beta_2 - \beta_1)(J_1^2 + J_2^2)}$$

$$J_{13} = 1 - (x \cos \varepsilon + \beta_1 y \sin \varepsilon)/r$$

$$J_{14} = (x \sin \varepsilon - \beta_1 y \cos \varepsilon)/r$$

The function $\psi_0'(z_2)$ is evaluated in the same way.

A number of individual stress analyses have been carried out to investigate how the size and shape of the zone of influence varies with the different parameters. Two plots of the relevant contours are shown in Figure 2.34. The data for Figure 2.34a are the following: $a = 1$, $b = 0.5$, $p_x = 0.5$, $p_y = 1$, $p_{xy} = 0$, $\beta_1 = 0.6589$ and $\beta_2 = 3.3077$. The values of β_1 and β_2 correspond to the situation where the planes of stratification are horizontal (y axis = axis of symmetry). The data for Figure 2.34b are the same, except that in this case the planes of stratification are vertical and as a consequence the values of β_1 and β_2 are replaced by their reciprocals, i.e. $\beta_1 = 1.5177$ and $\beta_2 = 0.3023$. It is instructive to compare these diagrams with Figure 2.28, which gives the stress contours for an elliptical hole in an isotropic rock for

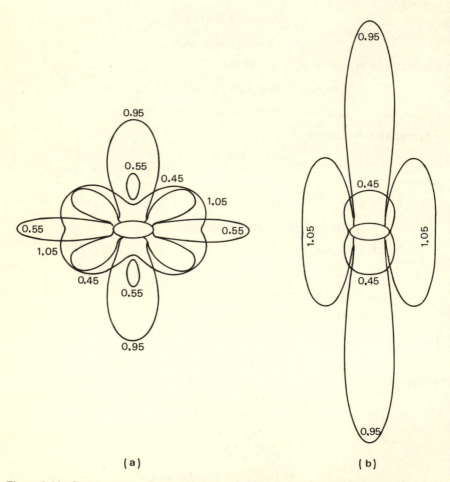

(a)   (b)

Figure 2.34 Stress contours for an elliptical excavation with a width:height ratio of 2 and principal field stresses of 1 vertical and 0.5 horizontal. The stress contours in (a) are for $\beta_1 = 0.6589$ and $\beta_2 = 3.3077$, and those in (b) are for $\beta_1 = 1.5177$ and $\beta_2 = 0.3023$.

the same width : height ratio and field stresses. It will be seen that in Figures 2.34a and b the zone of influence is stretched in the direction or stratification but that there is little overall change in the perpendicular direction.

Consideration will now be given to the restricted case where $p_{xy} = 0$ and the axis of elastic symmetry is either vertical or horizontal ($\rho = 0$ or $\pi/2$). As in the isotropic case the salient dimensions of the zone of influence are its width and height, as determined from the intercepts of the relevant stress contours on the x and y axes. Expressions for the stress components on these axes can be obtained from the results of the previous analysis by putting $\rho = 0$ or $\pi/2$ and $y = 0$ or $x = 0$, whichever alternative is relevant. Taking $\rho = 0$, the following results are obtained

(a) On the x axis:

$$\sigma_x = p_x - G_{x1}\left(1 - \frac{|x|}{X_1}\right) - G_{x2}\left(1 - \frac{|x|}{X_2}\right)$$

$$\sigma_y = p_y - G_{y1}\left(1 - \frac{|x|}{X_1}\right) - G_{y2}\left(1 - \frac{|x|}{X_2}\right)$$

(2.80)

(b) On the y axis:

$$\sigma_x = p_x - G_{x1}\left(1 - \frac{\beta_1|y|}{Y_1}\right) - G_{x2}\left(1 - \frac{\beta_2|y|}{Y_2}\right)$$

$$\sigma_y = p_y + G_{y1}\left(1 - \frac{\beta_1|y|}{Y_1}\right) + G_{y2}\left(1 - \frac{\beta_2|y|}{Y_2}\right)$$

(2.81)

where

$$G_{x1} = \frac{\beta_1^2(bp_x - \beta_2 a p_y)}{(\beta_2 - \beta_1)(a - \beta_1 b)}$$

$$G_{x2} = \frac{\beta_2^2(bp_x - \beta_1 a p_y)}{(\beta_1 - \beta_2)(a - \beta_2 b)}$$

$$G_{y1} = \frac{(bp_x - \beta_2 a p_y)}{(\beta_2 - \beta_1)(a - \beta_1 b)}$$

(2.82)

$$G_{y2} = \frac{(bp_x - \beta_1 a p_y)}{(\beta_1 - \beta_2)(a - \beta_2 b)}$$

$$X_1 = (x^2 + \beta_1^2 b^2 - a^2)^{1/2}$$

$$X_2 = (x^2 + \beta_2^2 b^2 - a^2)^{1/2}$$

$$Y_1 = (\beta_1^2 y^2 - \beta_1^2 b^2 + a^2)^{1/2}$$

(2.83)

$$Y_2 = (\beta_2^2 y^2 - \beta_2^2 b^2 + a^2)^{1/2}$$

These equations can also be used for the situation where $\rho = \pi/2$ by replacing β_1 and β_2 by their reciprocals.

The above equations may be used in a search routine to determine the width W_i and height H_i of the zone of influence. The results of such an investigation for 5% zones of influence are presented in Figures 2.35 and 2.36, both for $\rho = 0$, the former being for $\beta_1 = 0.6589$ and $\beta_2 = 3.3077$ and the latter for $\beta_1 = 0.6457$ and $\beta_2 = 4.648$. These diagrams should be compared with Figure 2.30 which gives the results for the isotropic case. As before, L is equal to the width or height of the excavation, whichever is the larger, and the value of q for each curve is shown at the sides of the figures.

Before bringing this chapter to a close, it is necessary to make some comment on the choice of the values of β_1 and β_2 used in the calculations. As I did not have a reliable set of test results to hand, and as an arbitrary choice could lead to misleading results, it was necessary to have recourse to a method of generating realistic values of β_1 and β_2.

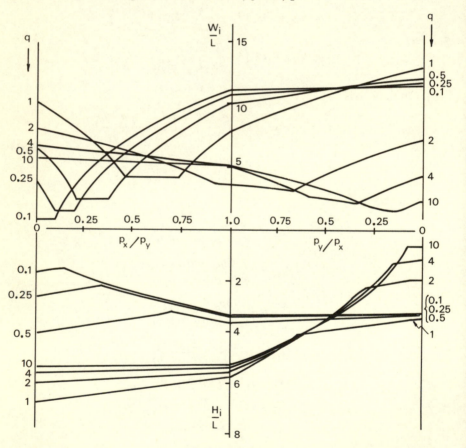

Figure 2.35 Diagram showing how the width W_i and height H_i of the zone of influence of an elliptical excavation in a transversely isotropic rock varies with the width : height ratio q of the excavation and the field stress ratio $p_x : p_y$ (or $p_y : p_x$). The curves are drawn for $\beta_1 = 0.6589$ and $\beta_2 = 3.3077$.

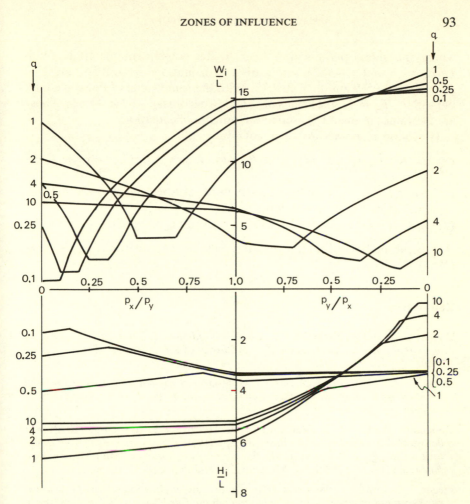

Figure 2.36 Diagram showing how the width W_i and height H_i of the zone of influence of an elliptical excavation in a transversely isotropic rock varies with the width : height ratio q of the excavation and the field stress ratio $p_x : p_y$ (or $p_y : p_x$). The curves are drawn for $\beta_1 = 0.6457$ and $\beta_2 = 4.648$.

I decided to use a simple model material consisting of alternate layers of two different isotropic materials, as shown in Figure 2.31. Assuming that the layers are bonded together perfectly so that there is no slip or separation, it is a fairly straightforward matter to work out the equivalent elastic properties of the composite material. Taking all the layers to be of equal thickness and endowed with the same Poisson's ratio, it can be shown that

$$E = \tfrac{1}{2}(E_1 + E_2)$$

$$\nu_0 = \nu = \nu_1 = \nu_2$$

$$N = 2G_1 G_2/(G_1 + G_2) \qquad\qquad (2.84)$$

$$n = (E_1 + E_2)^2/(4E_1 E_2)$$

where the subscripts 1 and 2 refer to the constituent materials. With $E_2/E_1 = 16$ and $\nu = 0.25$, application of Equations 2.84, 2.56, and 2.60 gives $\beta_1 = 0.6589$ and $\beta_2 = 3.3077$. With the same value of Poisson's ratio but with $E_2/E_1 = 32$, the values of β_1 and β_2 become 0.6457 and 4.648. These are the values of β_1 and β_2 used in the previous calculations.

The following points are worthy of note:

(a) Interchanging the values of β_1 and β_2 has no effect on the stress distribution.
(b) When $E_2/E_1 = 1$, the layered structure vanishes and $\beta_1 = \beta_2 = 1$ (the values for an isotropic material).
(c) As E_2/E_1 increases, $\beta \to 1/\sqrt{[2(1 + \nu)]}$ and $\beta_2 \to \sqrt{[E_2/2(1 - \nu)E_1]}$.

A general analysis of multilayer materials has been carried out by Eissa (1980).

References

Duvall, W. I. and W. Blake 1967. *Nonuniform radial loads applied to the boundary of a circular hole in an infinite plate.* US Bureau of Mines Report of Investigation 7030.

Eissa, E. A. 1980. *Stress analysis of underground excavations in isotropic and stratified rock using the boundary element method.* PhD thesis, University of London.

Goodman, R. E., T. K. Van and F. E. Heuzé 1972. Measurement of rock deformability in boreholes. In *Basic and applied rock mechanics*, Proc. 10th symp. rock mech., K. E. Gray (ed.), 523–55. New York: Society of Mining Engineers, American Institute of Mining, Metallurgical and Petroleum Engineers.

Jaeger, J. C. and N. G. W. Cook 1964. Theory and application of curved jacks for measurement of stresses. In *State of stress in the Earth's crust*, W. R. Judd (ed.), 381–95. New York: Elsevier.

Milne-Thomson, L. M. 1938. *Theoretical hydrodynamics.* London: Macmillan.
Mindlin, R. D. 1940. Stress distribution around a tunnel. *Trans Am. Soc. Civ. Engrs* **105**, 1117–53.

Savin, G. N. 1961. *Stress concentrations around holes.* London: Pergamon.

Timoshenko, S. P. and J. N. Goodier 1951. *Theory of elasticity.* New York: Wiley.

3 General two-dimensional slope stability analysis

E. HOEK

3.1 Introduction

The following analysis, originally published by Sarma (1979) and modified by me, is a general method of limit equilibrium analysis which can be used to determine the stability of slopes of a variety of shapes. Slopes with complex profiles with circular, non-circular, or planar sliding surfaces or any combination of these can be analysed using this method. In addition, active–passive wedge failures such as those which occur in spoil piles on sloping foundations or in clay core dam embankments can also be analysed. The analysis allows different shear strengths to be specified for each slice side and base. The freedom to change the inclination of the slice sides also allows the incorporation of specific structural features such as faults or bedding planes. External forces can be included for each slice, and submergence of any part of the slope is automatically incorporated into the analysis.

The geometry of the sliding mass is defined by the coordinates XT_i, YT_i, XB_i, YB_i, XT_{i+1}, YT_{i+1} and XB_{i+1}, YB_{i+1} of the corners of a number of three- or four-sided elements as shown in Figure 3.1. The phreatic surface is defined by the coordinates XW_i, YW_i and XW_{i+1}, YW_{i+1} of its intersection with the slice sides. A closed form solution is used to calculate the critical horizontal acceleration K_c required to induce a state of limiting equilibrium in the sliding mass. The static factor of safety F is then found by reducing the shear strength values $\tan \phi$ and c to $\tan \phi/F$ and c/F until the critical acceleration K_c is reduced to zero.

In order to determine whether the analysis is acceptable, a check is carried out to assess whether all the effective normal stresses acting across the bases and sides of the slices are positive. If negative stresses are found, the slice geometry or the groundwater conditions must be changed until the stresses are all positive. An additional check for moment equilibrium is described, but it has not been incorporated into the program listed at the end of the chapter because it involves a significant increase in computational effort and because it is seldom required for normal slope stability problems.

95

3.2   Geometrical calculations

The geometry of the ith slice is defined in Figure 3.1. Note that the value of the x coordinate should always increase from the toe of the slope. Assuming that ZW_i, δ_i and d_i are available from the previous slice:

$$d_{i+1} = [(XT_{i+1} - XB_{i+1})^2 + (YT_{i+1} - YB_{i+1})^2]^{1/2} \tag{3.1}$$

$$\delta_{i+1} = \arcsin [(XT_{i+1} - XB_{i+1})/d_{i+1}] \tag{3.2}$$

$$b_i = XB_{i+1} - XB_i \tag{3.3}$$

$$\alpha_i = \arctan [(YB_{i+1} - YB_i)/b_i] \tag{3.4}$$

$$W_i = \tfrac{1}{2}\gamma_r|(YB_i - YT_{i+1})(XT_i - XB_{i+1}) + (YT_i - YB_{i+1})(XT_{i+1} - XB_i)| \tag{3.5}$$

$$ZW_{i+1} = (YW_{i+1} - YB_{i+1}) \tag{3.6}$$

where γ_r is the unit weight of the material forming the slice and W_i is the weight of the slice.

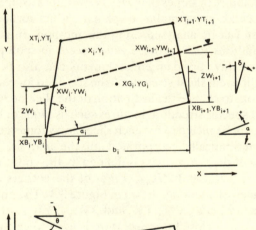

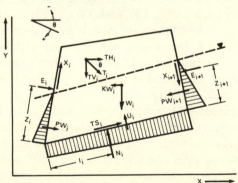

Figure 3.1 Definition of geometry and forces acting on the ith slice.

3.3　Calculation of water forces

In order to cover all possible groundwater conditions, including submergence of any part of the slope, the four cases defined in Figure 3.2 have to be considered. In all cases, the uplift U_i acting on the base of the slice is given by

$$U_i = \tfrac{1}{2}\gamma_{\mathrm{w}} |(YW_i - YB_i + YW_{i+1} - YB_{i+1})b_i/\cos \alpha_i| \qquad (3.7)$$

where γ_{w} is the unit weight of water.

Case 1: no submergence of slice

$$YT_i > YW_i \quad \text{and} \quad YT_{i+1} > YW_{i+1}$$

$$PW_i = \tfrac{1}{2}\gamma_{\mathrm{w}} |(YW_i - YB_i)^2/\cos \delta_i| \qquad (3.8)$$

$$PW_{i+1} = \tfrac{1}{2}\gamma_{\mathrm{w}} |(YW_{i+1} - YB_{i+1})^2/\cos \delta_{i+1}| \qquad (3.9)$$

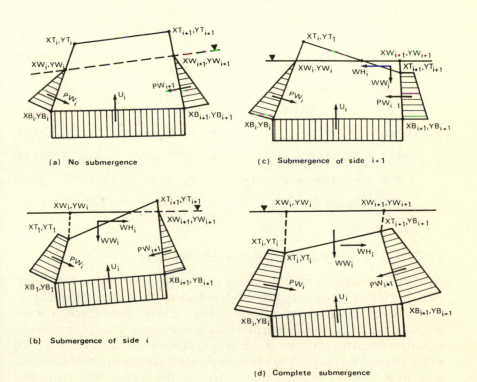

(a) No submergence

(c) Submergence of side i + 1

(b) Submergence of side i

(d) Complete submergence

Figure 3.2　Definition of water forces.

Case 2: submergence of side i only

$$YT_i < YW_i \quad\text{and}\quad YT_{i+1} > YW_{i+1}$$

$$PW_i = \tfrac{1}{2}\gamma_w|(2YW_i - YT_i - YB_i)(YT_i - YB_i)/\cos \delta_i| \tag{3.10}$$

$$PW_{i+1} = \tfrac{1}{2}\gamma_w|(YW_{i+1} - YB_{i+1})^2/\cos \delta_{i+1}| \tag{3.11}$$

$$WW_i = \tfrac{1}{2}\gamma_w|(YW_i - YT_i)^2(XT_{i+1} - XT_i)/(YT_{i+1} - YT_i)| \tag{3.12}$$

$$WH_i = \tfrac{1}{2}\gamma_w(YW_i - YT_i)^2 \tag{3.13}$$

where WW_i and WH_i are the vertical and horizontal forces applied to the surface of the slice as a result of the submergence of part of the slice. Note that the horizontal force WH_i acts in a positive direction when $YT_{i+1} > YT_i$ and in a negative direction when $YT_{i+1} < YT_i$.

Case 3: submergence of side i + 1 only

$$YT_i > YW_i \quad\text{and}\quad YT_{i+1} < YW_{i+1}$$

$$PW_i = \tfrac{1}{2}\gamma_w|(YW_i - YB_i)^2/\cos \delta_i| \tag{3.14}$$

$$PW_{i+1} = \tfrac{1}{2}\gamma_w|(2YW_{i+1} - YT_{i+1} - YB_{i+1})(YT_{i+1} - YB_{i+1})/\cos \delta_{i+1}| \tag{3.15}$$

$$WW_i = \tfrac{1}{2}\gamma_w|(YW_{i+1} - YT_{i+1})^2(XT_{i+1} - XT_i)/(YT_{i+1} - YT_i)| \tag{3.16}$$

$$WH_i = \tfrac{1}{2}\gamma_w(YW_{i+1} - YT_{i+1})^2 \tag{3.17}$$

Case 4: complete submergence of slice i

$$XT_i < XW_i \quad\text{and}\quad XT_{i+1} < XW_{i+1}$$

$$PW_i = \tfrac{1}{2}\gamma_w|(2YW_i - YT_i - YB_i)(YT_i - YB_i)/\cos \delta_i| \tag{3.18}$$

$$PW_{i+1} = \tfrac{1}{2}\gamma_w|(2YW_{i+1} - YT_{i+1} - YB_{i+1})(YT_{i+1} - YB_{i+1})/\cos \delta_{i+1}| \tag{3.19}$$

$$WW_i = \tfrac{1}{2}\gamma_w|(YW_i - YT_i + YW_{i+1} - YT_{i+1})(XT_{i+1} - XT_i)| \tag{3.20}$$

$$WH_i = \tfrac{1}{2}\gamma_w|(YW_i - YT_i + YW_{i+1} - YT_{i+1})(YT_{i+1} - YT_i)| \tag{3.21}$$

3.3.1   Water forces on the first and last slice sides

Although the water force PW_i acting on the first slice side and the force PW_{n+1} acting on the $(n + 1)$th slice side (which could be a tension crack) are calculated by means of the equations listed above, these forces are not normally used in the calculation of critical acceleration. These forces can be important when the slope toe is submerged or when a tension crack is filled with water and the simplest way to incorporate these forces into the analysis is to treat them as external forces. Hence, the vertical and horizontal components of these forces are given by

$$TV_1 = PW_1 \sin \delta_1 \tag{3.22}$$

$$TH_1 = PW_1 \cos \delta_1 \qquad (3.23)$$

$$TV_n = PW_{n+1} \sin \delta_{n+1} \qquad (3.24)$$

$$TH_n = PW_{n+1} \cos \delta_{n+1} \qquad (3.25)$$

where n is the total number of slices included in the analysis.

3.4 Calculation of critical acceleration K_c

The critical acceleration K_c required to bring the slope to a condition of limiting equilibrium is given by

$$K_c = AE/PE \qquad (3.26)$$

where

$$AE = a_n + a_{n-1}e_n + a_{n-2}e_ne_{n-1} + \cdots + a_1e_ne_{n-1}\cdots e_3e_2 \qquad (3.27)$$

$$PE = p_n + p_{n-1}e_n + p_{n-2}e_ne_{n-1} + \cdots + p_1e_ne_{n-1}\cdots e_3e_2 \qquad (3.28)$$

$$a_i = Q_i[(W_i + TV_i) \sin (\phi_{Bi} - \alpha_i) - TH_i \cos (\phi_{Bi} - \alpha_i) + R_i \cos \phi_{Bi}$$
$$+ S_{i+1} \sin (\phi_{Bi} - \alpha_i - \delta_{i+1}) - S_i \sin (\phi_{Bi} - \alpha_i - \delta_i)] \qquad (3.29)$$

$$p_i = Q_iW_i \cos (\phi_{Bi} - \alpha_i) \qquad (3.30)$$

$$e_i = Q_i[\cos (\phi_{Bi} - \alpha_i + \phi_{Si} - \delta_i)/\cos \phi_{Si}] \qquad (3.31)$$

$$Q_i = \cos \phi_{Si+1}/\cos (\phi_{Bi} - \alpha_i + \phi_{Si+1} - \delta_{i+1}) \qquad (3.32)$$

$$S_i = c_{Si}d_i - PW_i \tan \phi_{Si} \qquad (3.33)$$

$$S_{i+1} = c_{Si+1}d_{i+1} - PW_{i+1} \tan \phi_{Si+1} \qquad (3.34)$$

$$R_i = c_{Bi}b_i/\cos \alpha_i - U_i \tan \phi_{Bi} \qquad (3.35)$$

3.5 Calculation of factor of safety

For slopes when the critical acceleration K_c is not equal to zero, the static factor of safety is calculated by reducing the shear strength simultaneously on all sliding surfaces until the acceleration K_c, calculated by means of Equation 3.26, reduces to zero. This is achieved by substitution in Equations 3.29 to 3.35 of the following shear strength values

$$c_{Bi}/F, \tan \phi_{Bi}/F, c_{Si}/F, \tan \phi_{Si}/F, c_{Si+1}/F \text{ and } \tan \phi_{Si+1}/F$$

3.6 Check on acceptability of solution

Having determined the value of K for a given factor of safety, the forces acting on the sides and base of each slice are found by progressive solution of the following equations, starting from the known condition that $E_1 = 0$.

$$E_{i+1} = a_i - p_i K + E_i e_i \tag{3.36}$$

$$X_i = (E_i - PW_i) \tan \phi_{Si} + c_{Si} d_i \tag{3.37}$$

$$N_i = (W_i + TV_i + X_{i+1} \cos \delta_{i+1} + X_i \cos \delta_i - E_{i+1} \sin \delta_{i+1} + E_i \sin \delta_i$$
$$+ U_i \tan \phi_{Bi} \sin \alpha_i - c_{Bi} b_i \tan \alpha_i) \cos \phi_{Bi} / \cos (\phi_{Bi} - \alpha_i) \tag{3.38}$$

$$TS_i = (N_i - U_i) \tan \phi_{Bi} + c_{Bi} b_i / \cos \alpha_i \tag{3.39}$$

The effective normal stresses acting across the base and the sides of a slice are calculated as follows:

$$\sigma'_{Bi} = (N_i - U_i) \cos \alpha_i / b_i \tag{3.40}$$

$$\sigma'_{Si} = (E_i - PW_i) / d_i \tag{3.41}$$

$$\sigma'_{Si+1} = (E_{i+1} - PW_{i+1}) / d_{i+1} \tag{3.42}$$

In order for the solution to be acceptable, all effective normal stresses must be positive.

A final check to determine whether moment equilibrium conditions are satisfied for each slice is recommended by Sarma (1979). Referring to Figure 3.1 and taking moments about the lower left-hand corner of the slice:

$$N_i l_i - X_{i+1} b_i \cos (\alpha_i + \delta_{i+1}) / \cos \alpha_i - E_i Z_i$$
$$+ E_{i+1}[Z_{i+1} + b_i \sin (\alpha_i + \delta_{i+1}) / \cos \alpha_i)]$$
$$- W_i(XG_i - X_{Bi}) + K_c W_i(YG_i - Y_{Bi}) - TV_i(X_i - XG_i)$$
$$+ TH_i(Y_i - YG_i) = 0 \tag{3.43}$$

where XG_i, YG_i are the coordinates of the centre of gravity of the slice and X_i, Y_i are the coordinates of the point of action of the force T_i.

Starting from the first slice at the toe of the slope, where $Z_i = 0$, assuming a value of l_i, the moment arm Z_{i+1} can be calculated or vice versa. The values of Z_i and Z_{i+1} should lie within the slice boundary, preferably in the middle third.

3.7   Computer solution for Sarma analysis

A listing of a computer program for the analysis presented above is given in Appendix A at the end of this chapter. This program has been written in the simplest form of BASIC, and great care has been taken to ensure that there are no machine-dependent commands in the program. Hence, it should be possible to key this program into any computer which runs Microsoft or equivalent BASIC and to modify it for any other form of BASIC. The program has also been carefully prepared so that it can be compiled into machine language using a BASIC compiler. The compiled program will run about six times faster than the program listed in Appendix A.

A graphics option is built into the program which allows the user to view the geometry of the slope being analysed. This option assumes that BASICA or an equivalent form of BASIC which supports graphics is available and that the computer has IBM-compatible graphics capability. If these facilities are not available, the program will operate correctly but the display will be incomplete.

A critical component of the program is the factor of safety iteration in which the shear strength values are progressively reduced (or increased) until the static factor of safety (for $K = 0$) is found. Experience has shown that this iteration can be a very troublesome process and that severe numerical instability can occur if inappropriate values of F are used. The iteration technique used in the listed program is described below.

Figure 3.3 gives a plot of factor of safety F versus acceleration K for a range of friction angles for a typical slope analysis. This plot reveals that the curve

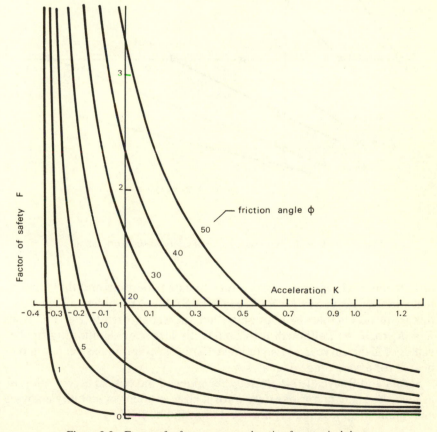

Figure 3.3   Factor of safety versus acceleration for a typical slope.

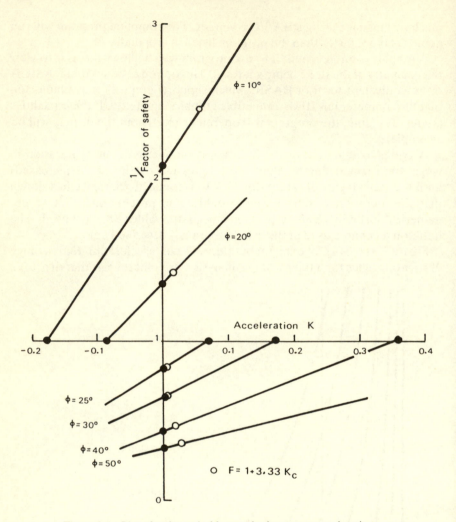

Figure 3.4 Plot of reciprocal of factor of safety versus acceleration.

of F versus K closely resembles a rectangular hyperbola and this suggests that a plot of $1/F$ versus K should be a straight line. As shown in Figure 3.4, this is an acceptable assumption within the range of interest, i.e. from $K = K_c$ to $K = 0$, although for significantly larger or significantly smaller values of K the curve is no longer linear. This observation has been found to be true for a wide range of analyses.

Sarma and Bhave (1974) plotted the values of the critical acceleration K_c against the static factor of safety F for a large number of stability analyses and found an approximately linear relationship defined by

$$F = 1 + 3.33K_c \qquad (3.44)$$

Although this relationship does not provide sufficient accuracy for the very wide range of problems encountered when applying this analysis to rock mechanics problems, it does give a useful point close to the $K = 0$ axis on the plot of $1/F$ versus K, as shown in Figure 3.4. A linear interpolation or extrapolation, using this point and the value of K_c (at $F = 1$), gives an accurate estimate of the static factor of safety. This technique has proved to be very fast and efficient and has been incorporated into the program. For critical cases, in which it is considered essential to plot the complete F versus K curve, an optional subroutine has been provided in the program to enable the user to produce such a plot.

3.8   Problems with negative stresses

The effective normal stresses across the sides and base of each slice are calculated by means of Equations 3.40 to 3.42 and, in order for the solution to be acceptable, these stresses must all be positive. The reasons for the occurrence of negative stresses and some suggested remedies are discussed below.

Negative stresses can occur near the top of a slope when the lower portion of the slope is less stable and hence tends to slide away from the upper portion of the slope. This is the condition which leads to the formation of tension cracks in actual slopes, and the negative stresses in the numerical solution can generally be eliminated by placing a tension crack at an appropriate position in the slope.

Negative stresses at the toe of a slope are sometimes caused by an excessively strong toe. This can occur when the upward curvature of a deep-seated failure surface becomes too severe in the toe region. Flattening the curvature or reducing the shear strength along the base will generally solve this problem.

Excessive water pressures within the slope can give rise to negative stresses, particularly near the top of the slope where the normal stresses are low. Reducing the level of the phreatic surface in the region in which negative stresses occur will usually eliminate these negative stresses.

Inappropriate selection of the slice geometry, particularly the inclination of the slice sides, can give rise to negative stress problems. This is an important consideration in rock mechanics when pre-existing failure surfaces such as joints and faults are included in the analysis. If a potential failure path with a lower shear strength than that of the pre-existing surface exists in the sliding mass, negative stresses can occur along the pre-existing surface which has been chosen as a slice side. Sarma (1979) has shown that the most critical slice side inclinations are approximately normal to the basal failure surface. In the case of a circular failure in homogeneous soil, these slide sides are approximately radial to the centre of curvature of the failure surface.

A rough or irregular failure surface can also give rise to negative stress problems if it causes part of the sliding mass to be significantly more stable than an adjacent part. During the early stages of development of this program, I compared answers against solutions for the same problem obtained from Bishop circular failure analyses. It was found that in order to obtain absolute agreement between the solutions the coordinates of the failure surface had to be calculated to ensure that the slice base inclinations were identical in the two analyses. Consequently, for critical problems, reading the slice base coordinates from a drawing may not be adequate and it may be necessary to calculate these coordinates to ensure that undulations are not built into the analysis.

3.9   Drainage of slopes

Three options for analysing the influence of drainage upon the stability of slopes have been included in the program listed in Appendix A.

The first option involves inserting a value for zero for the unit weight of water during initial entry of the data. This will activate an automatic routine in the program which will set all water forces to zero and give the solution for a fully drained slope.

The second option provides the user with the facility for changing the unit weight of water during the operation of the program. This results in a pore-pressure ratio (r_u) type of analysis such as that commonly used in soil mechanics (see Bishop & Morgenstern 1960). This analysis is useful for sensitivity studies on the influence of drainage on slope stability since it provides the user with a very fast means of changing the water pressures throughout the slope.

The final method of analysing the influence of drainage is to change the phreatic surface coordinates on each slice boundary. This method is rather tedious but it probably represents the actual field conditions more realistically than the r_u analysis described above.

The water-pressure distributions assumed in this analysis are illustrated in Figures 3.1 and 3.2. These distributions are considered to be representative of those most commonly occurring in the field. There are, however, situations in which these water-pressure distributions are inappropriate. The best example of such a situation is a dam foundation in which the water-pressure distribution is modified by the presence of grout and drainage curtains. The simplest way to account for such changes in the analysis presented here is to calculate the change in total uplift force on the base of each slice influenced by drainage and grouting and then to apply this change as a stabilizing external force acting normal to the slice base.

3.10   Incorporation of non-linear failure criteria

Hoek (1983) has discussed the question of non-linear failure criteria for heavily jointed rock masses and has given an example of the analysis of a large open pit mine slope in such materials. Since the Sarma analysis calculates the effective normal stresses on each slice side and base, these values can be used to determine the instantaneous cohesion and friction angle acting on these surfaces. An iterative technique is used to change these shear strength values until the difference between factors of safety calculated in successive iterations is acceptably small. Three or four iterations are usually sufficient to give an acceptable answer.

The iterative process described above is relatively easy to build into the analysis presented in Appendix A, but in the interests of space this has not been done in this chapter. In addition, non-linear analyses are generally only carried out for fairly complex problems and only after a large number of sensitivity studies using linear failure criteria have been performed. In such cases, the user is generally seeking a fundamental understanding of the mechanics of the slope behaviour, and it is advantageous to carry out the non-linear analysis manually in order to enhance this understanding.

3.11   Recommended steps in carrying out an analysis

When applying this analysis to an actual slope problem a great deal of time can be wasted if too detailed an analysis is attempted at the beginning of the study.

The first step in any analysis involves a determination of the most critical failure surface. Except where this surface has been clearly pre-defined by existing geological weakness planes or an observed failure surface, some form of search for the critical failure surface must be carried out. A good starting point for such a search is a set of charts such as those devised by Hoek and Bray (1981) and reproduced in Figures 3.5 and 3.6. These give a first estimate of the location of the centre of rotation of a critical circle and the position of a tension crack in a homogeneous slope.

Based upon some estimate or educated guess of the critical failure surface location, the sliding mass is divided into slices, using the smallest possible number of slices to approximate the geometry. Usually three or four slices will suffice at this stage since only a very crude analysis is required to check the critical failure geometry.

A number of trial analyses with different failure surface locations should then be carried out. It will be found that a set of critical conditions will quickly be found and a more refined model can then be constructed. Unless the slope geometry is extremely complex, five to ten slices will generally be found to give an acceptable level of accuracy for this refined analysis.

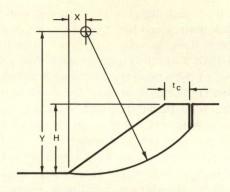

Distance X from slope toe

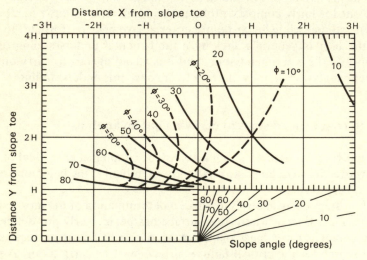

Distance Y from slope toe

Slope angle (degrees)

Angle of slope face (degrees)

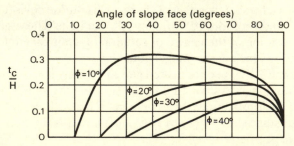

Figure 3.5 Approximate locations of the centre of curvature of a circular failure surface and a tension crack in a drained homogeneous slope.

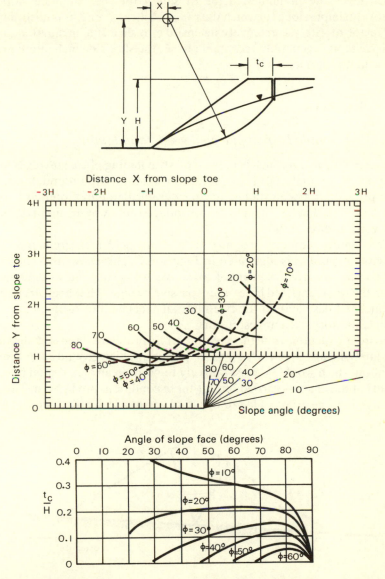

Figure 3.6 Approximate locations of the centre of curvature of a circular failure surface and a tension crack in a homogeneous slope with groundwater present.

Sarma (1979) suggests that the optimum inclination of the slice sides should be determined by varying the inclination of each slice side while keeping the others fixed. The optimum inclination for each slice side is that which gives the minimum factor of safety for the complete slope. In relatively simple slopes in which the range of shear strengths is fairly limited, the factor of safety is relatively insensitive to slice side inclination and, in such cases, it is generally acceptable to set the slice side inclination normal to the failure surface.

3.12 Examples

3.12.1 Example 1: spoil pile on a weak foundation

A common problem which occurs in the strip mining of coal involves failure of spoil (waste material) piles placed on weak inclined foundations. One such problem has been studied in considerable detail by Coulthard (1979), and the results that he obtained are reproduced below by means of the Sarma non-vertical slice method.

The failure geometry, reconstructed from field measurements by the CSIRO, Australia, is illustrated in Figure 3.7. This shows that the failure involves downward movement of an 'active' wedge and outward movement along the weak, inclined base of the 'passive' wedge. The shear strengths on the inclined base and on the two internal shear failure surfaces are based upon laboratory strength test results.

In order to demonstrate the negative stress problems discussed above, the first analysis carried out is for a waste pile with a high groundwater table – a situation which would be most unlikely to occur in an actual spoil pile. The results of the analysis carried out for these conditions are listed in Table 3.1 and a plot of the graph of factor of safety versus acceleration is given by the

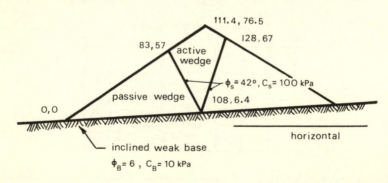

Figure 3.7 Geometry of a spoil pile on a weak foundation analysed by Coulthard (1979).

Table 3.1 Analysis of the stability of a spoil pile on a weak, inclined foundation (Fig. 3.7): Sarma non-vertical slice analysis.

Analysis no. 1: spoil pile on a weak foundation
Unit weight of water = 10

Side number	1	2	3
coordinate XT	0.00	83.00	111.40
coordinate YT	0.00	57.00	76.50
coordinate XW	0.00	83.00	128.00
coordinate YW	0.00	57.00	67.00
coordinate XB	0.00	108.00	128.00
coordinate YB	0.00	6.40	67.00
friction angle	0.00	42.00	0.00
cohesion	0.00	100.00	0.00

Slice number	1	2
rock unit weight	15.70	15.70
friction angle	6.00	42.00
cohesion	10.00	100.00
force T	0.00	0.00
angle θ	0.00	0.00

Effective normal stresses		
base	147.15	−14.13
side	0.00	−130.64

Acceleration K_c = −0.2087
Factor of safety = 0.26
Negative effective normal stresses – solution unacceptable

dashed line in Figure 3.8. Note that negative stresses occur on both internal shearing surfaces in this analysis and the calculated factor of safety of 0.26 is unacceptable. It is also important to note that the plot of F versus K has an asymptote of $F = 0.428$ and that factors of safety of less than this value are meaningless.

This example demonstrates that, under certain circumstances, the iteration technique used in the program (and all other iteration techniques tried during the development of the program) will choose the incorrect solution. As shown in Figure 3.8, the factor of safety for $K = 0$ is 0.48, whereas the second root of 0.26 has been chosen by the iteration technique. Fortunately, this problem is very rare and, as far as I have been able to ascertain, is always associated with negative stresses which generate the message that the solution is unacceptable. Nevertheless, for critical problems, it is recommended that the curve of factor of safety versus acceleration be plotted out to ensure that the correct solution has been chosen.

The second solution is for a drained spoil pile. A factor of safety of 1.20 is obtained for this case and all the effective normal stresses are positive. The plot of factor of safety versus acceleration, given as the solid line in Figure

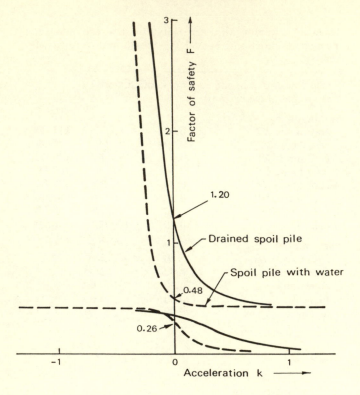

Figure 3.8 Plot of factor of safety versus acceleration K for a spoil pile on an inclined weak foundation.

3.8, shows that the value of 1.20 is well above the asymptote of 0.428 which, interestingly, is the same as for the previous analysis.

A third analysis, using the same geometry as illustrated in Figure 3.7 but with the slope drained and the cohesion on side 2 reduced to zero, produces a factor of safety of 1.00. This limiting equilibrium condition is identical to that obtained by Coulthard (1979) using a two-wedge analysis similar to that proposed by Seed and Sultan (1967) for sloping-core embankment dams.

3.12.2   Example 2: open pit coal mine slope

Figure 3.9 illustrates the geometry of a slope problem in a large open pit coalmine. A thin coal seam is overlain by soft tuff and existing failures in the slope show that sliding occurs along the coal seam with the toe breaking out through the soft tuff. In the case illustrated, a reservoir close to the crest of the slope recharges the slope with water and results in the high groundwater surface illustrated. Laboratory tests and back-analysis of previous failures give a friction angle of 18° and a cohesion of zero along the coal seam and a friction angle of 30° and cohesion of 2 t/m² for failure through the soft tuff. The unit weight of the tuff is 2.1 t/m³ and the unit weight of water is 1.0 t/m³.

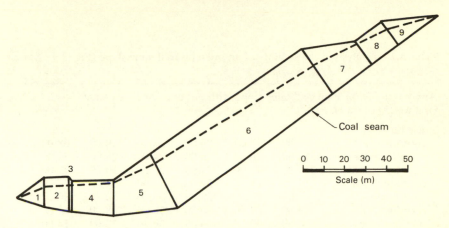

Figure 3.9   Slope geometry for an open pit coal mine slope.

A printout produced by the program listed in Appendix A is reproduced in Table 3.2. This shows that the factor of safety for the slope illustrated in Figure 3.9 is 1.17. A sensitivity study of drainage shows that 50% drainage (reducing the unit weight of water to 0.5 t/m^3) increases the factor of safety to 1.41 and that complete drainage of the slope gives a factor of safety of 1.65. In the actual case upon which this example is based, the reservoir above the slope was drained, but in adjacent slopes long horizontal drain holes were used to reduce the water pressures in the slopes.

3.12.3   Example 3: partially submerged rockfill slope

Figure 3.10 illustrates the geometry of a rockfill slope placed under water onto a sandy river bottom. The rockfill is partially submerged, and the

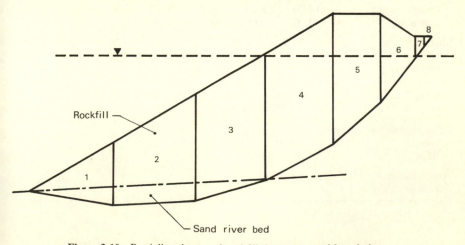

Figure 3.10   Partially submerged rockfill slope on a sand foundation.

Table 3.2 Analysis of the stability of an open pit coal mine slope (Fig. 3.9): Sarma non-vertical slice analysis.

Analysis no. 2: open pit coal mine slope with tuff overlying coal seam
Unit weight of water = 1

Side number	1	2	3	4	5	6
coordinate XT	4.00	17.00	29.00	30.00	50.00	68.00
coordinate YT	17.00	26.00	26.00	24.00	25.00	37.00
coordinate XW	4.00	17.00	29.00	30.00	50.00	70.00
coordinate YW	17.00	23.00	22.00	22.00	24.00	33.00
coordinate XB	4.00	17.00	29.00	30.00	50.00	80.00
coordinate YB	17.00	12.00	10.00	10.00	8.00	11.00
friction angle	0.00	30.00	30.00	30.00	30.00	18.00
cohesion	0.00	2.00	2.00	2.00	2.00	0.00

Slice number	1	2	3	4	5	6
rock unit weight	2.10	2.10	2.10	2.10	2.10	2.10
friction angle	30.00	30.00	30.00	30.00	30.00	30.00
cohesion	2.00	2.00	2.00	2.00	2.00	2.00
force T	0.00	0.00	0.00	0.00	0.00	0.00
angle θ	0.00	0.00	0.00	0.00	0.00	0.00

Effective normal stresses

base	29.91	37.07	22.52	29.22	44.54	27.92
side	0.00	23.69	41.25	48.08	60.63	63.72

Side number	7	8	9	10
coordinate XT	140.00	165.00	178.00	204.00
coordinate YT	88.00	90.00	99.00	103.00
coordinate XW	146.00	166.00	180.00	204.00
coordinate YW	80.00	89.00	96.00	103.00
coordinate XB	155.00	173.00	186.00	204.00
coordinate YB	65.00	80.00	89.00	103.00
friction angle	18.00	18.00	18.00	0.00
cohesion	0.00	0.00	0.00	0.00

Slice number	7	8	9
rock unit weight	2.10	2.10	2.10
friction angle	30.00	30.00	30.00
cohesion	2.00	2.00	2.00
force T	0.00	0.00	0.00
angle θ	`0.00	0.00	0.00

Effective normal stresses

base	18.89	11.90	6.64
side	14.41	11.11	3.25

Acceleration K_c = 0.1008
Factor of safety = 1.17

Table 3.3 Analysis of the stability of a partially submerged rockfill on a sand foundation (Fig. 3.10): Sarma non-vertical slice analysis.

Analysis no. 3: partially submerged rockfill slope on a sand foundation
Unit weight of water = 9.8

Side number	1	2	3	4	5	6
coordinate XT	2.30	10.00	18.00	24.20	31.00	35.50
coordinate YT	2.00	6.60	11.30	15.00	19.00	19.00
coordinate XW	2.30	10.00	18.00	24.20	31.00	35.50
coordinate YW	15.00	15.00	15.00	15.00	15.00	15.00
coordinate XB	2.30	10.00	18.00	24.20	31.00	35.50
coordinate YB	2.00	0.50	1.00	3.20	6.70	10.50
friction angle	0.00	35.00	40.00	40.00	40.00	40.00
cohesion	0.00	0.00	0.00	0.00	0.00	0.00

Slice number	1	2	3	4	5	6
rock unit weight	19.70	19.70	19.70	19.70	19.70	19.70
friction angle	33.00	33.00	33.00	40.00	40.00	40.00
cohesion	0.00	0.00	0.00	0.00	0.00	0.00
force T	0.00	0.00	0.00	0.00	0.00	0.00
angle θ	0.00	0.00	0.00	0.00	0.00	0.00

Effective normal stresses						
base	40.96	91.97	96.99	110.22	91.68	41.96
side	0.00	27.62	36.13	30.73	24.83	16.23

Side number	7	8	9
coordinate XT	39.00	40.00	40.40
coordinate YT	17.00	17.00	17.00
coordinate XW	39.00	40.00	40.40
coordinate YW	15.00	16.20	17.00
coordinate XB	39.00	40.00	40.40
coordinate YB	15.00	16.20	17.00
friction angle	40.00	33.00	0.00
cohesion	0.00	0.00	0.00

Slice number	7	8
rock unit weight	19.70	19.70
friction angle	40.00	33.00
cohesion	0.00	0.00
force T	0.00	0.00
angle θ	0.00	0.00

Effective normal stresses		
base	14.67	3.51
side	6.75	2.91

Acceleration K_c = 0.2106
Factor of safety = 1.91

failure surface (determined by a critical failure search using a conventional vertical slice analysis) involves both the rockfill and the sand base. A printout of the slope geometry, material properties, and calculated factor of safety is given in Table 3.3.

It is interesting to note that the critical acceleration for this slope is 0.2106. This means that for a pseudo-static analysis of earthquake loading a horizontal acceleration of $0.21g$ would be required to induce failure in the slope. Factors of safety corresponding to different pseudo-static horizontal acceleration levels can be found by using the optional subroutine included in the program to produce a plot similar to that illustrated in Figure 3.8. In the actual case upon which this example was based, a more complete dynamic earthquake analysis was performed, but in many cases a pseudo-static check on stability under earthquake loading is acceptable.

References

Bishop, A. W. and N. R. Morgenstern 1960. Stability coefficients for earth slopes. *Géotechnique* **10**(4), 129–50.

Coulthard, M. A. 1979. *Back-analysis of observed spoil failures.* Technical Report No. 83, Division of Applied Geomechanics, Commonwealth Science and Industrial Research Organization, Melbourne, Australia.

Hoek, E. 1983. Strength of jointed rock masses. *Géotechnique* **33**(3), 187–223.

Hoek, E. and J. W. Bray 1981. *Rock slope engineering*, 3rd edn. London: Institution of Mining and Metallurgy.

Sarma, S. K. 1979. Stability analysis of embankments and slopes. *J. Geotech. Engng Div., Am. Soc. Civ. Engrs* **105**(GT12), 1511–24.

Sarma, S. K. and M. V. Bhave 1974. Critical acceleration versus static factor of safety in stability analysis of earth dams and embankments. *Géotechnique* **24**(4), 661–5.

Seed, H. B. and H. A. Sultan 1967. Stability analyses for a sloping core embankment. *J. Soil Mech. Foundns Div., Am. Soc. Civ. Engrs* **93**(SM4) 45–67.

Appendix A   Basic computer program for Sarma non-vertical slice analysis

The BASIC program listed on pages 117–28 has been written for use on microcomputers which run Microsoft or equivalent BASIC. In order to utilize the graphics option it is necessary to use BASICA or an equivalent BASIC which supports graphics commands and to run the program on a computer fitted with an IBM or compatible graphics card. If no graphics facilities are available, the program will operate correctly but the graphics display will be incomplete.

The program can be compiled using a BASIC compiler; this produces a machine language program which runs about six times faster than the BASIC program. The compiled program may not drive printers fitted with serial interfaces (RS232C) and the printer manual should be consulted for instructions on initializing the printer. If it proves impossible to drive the printer from the compiled program, it will be found that the BASIC program will drive almost any printer.

The function keys for the main program execute the following operations:

- F1, print: prints the tabulated data and calculates critical acceleration and factor of safety as shown in Tables 3.1 to 3.3.
- F2, calculate: recalculates the critical acceleration and the factor of safety using the displayed data.
- F3, fos vs k: activates a subroutine to calculate values of the acceleration K for different factors of safety. A new display is used for this subroutine.
- F4, drain: enables the user to edit the line which displays the unit weight of water. This is used to change the unit weight of water in sensitivity studies using a pore-pressure ratio (ru) approach.
- F5, file: displays the data files already stored on the disk and requests a new file name. Stores the displayed data on a disk file.
- F6, restart: returns to the first page display which asks various questions before the data array is displayed.
- F7, quit: exits the program and returns to BASIC. The program may be reactivated by tying RUN and pressing enter.
- F8, view: activates the graphics display if a suitable graphics card is fitted.

Table A1 overleaf gives the definitions of the elements of the array $A(J, K)$ used in the program.

Table A1   Definition of array $A(J, K)$ for Sarma analysis.

Side number	1		k		$k + 1$
coordinate XT	$A(1,1)$		$A(1,k)$		$A(1,k+1)$
coordinate YT	$A(2,1)$		$A(2,k)$		$A(2,k+1)$
coordinate XW	$A(3,1)$		$A(3,k)$		$A(3,k+1)$
coordinate YW	$A(4,1)$		$A(4,k)$		$A(4,k+1)$
coordinate XB	$A(5,1)$		$A(5,k)$		$A(5,k+1)$
coordinate YB	$A(6,1)$		$A(6,k)$		$A(6,k+1)$
friction angle ϕ_B	$A(7,1)$		$A(7,k)$		$A(7,k+1)$
cohesion c_B	$A(8,1)$		$A(8,k)$		$A(8,k+1)$
Slice number		1		k	
rock unit weight γ_r		$A(10,1)$		$A(10,k)$	
friction angle ϕ_s		$A(11,1)$		$A(11,k)$	
cohesion c_s		$A(12,1)$		$A(12,k)$	
force T		$A(15,1)$		$A(15,k)$	
angle θ		$A(16,1)$		$A(16,k)$	
side length d	$A(17,1)$		$A(17,k)$		$A(17,k+1)$
side angle δ	$A(18,1)$		$A(18,k)$		$A(18,k+1)$
base length b		$A(19,1)$		$A(19,k)$	
base angle α		$A(20,1)$		$A(20,k)$	
slice weight W		$A(21,1)$		$A(21,k)$	
uplift force U		$A(22,1)$		$A(22,k)$	
water force P_W	$A(23,1)$		$A(23,k)$		$A(23,k+1)$
base cohesion c_B/F		$A(24,1)$		$A(24,k)$	
base friction $\tan \phi_B/F$		$A(25,1)$		$A(25,k)$	
side cohesion c_B/F	$A(26,1)$		$A(26,k)$		$A(26,k+1)$
side friction $\tan \phi_s/F$	$A(27,1)$		$A(27,k)$		$A(27,k+1)$
calculated R		$A(28,1)$		$A(28,k)$	
calculated S	$A(29,1)$		$A(29,k)$		$A(29,k+1)$
calculated Q		$A(30,1)$		$A(30,k)$	
calculated e		$A(31,1)$		$A(31,k)$	
calculated p		$A(32,1)$		$A(32,k)$	
calculated a		$A(33,1)$		$A(33,k)$	
calculated E	$A(34,1)$		$A(34,k)$		$A(34,k+1)$
calculated X		$A(35,1)$		$A(35,k)$	
calculated N		$A(36,1)$		$A(36,k)$	
calculated T_S		$A(37,1)$		$A(37,k)$	
calculated σ_b		$A(38,1)$		$A(38,k)$	
calculated σ_s	$A(39,1)$		$A(39,k)$		$A(39,k+1)$

```
10 ' SARMA - NON-VERTICAL SLICE METHOD OF SLOPE STABILITY ANALYSIS
20 ' Version 2.0:Written by Dr.E.Hoek, Golder Associates, April 1986
30 ' Reference : Sarma, S.K. (1979), Stability analysis of embankments
40 '             and slopes, J. Geotech. Engg. Div., ASCE., Vol. 105,
50 '             No. GT12, pages 1511-1524.
60 '
70 ' Dimensioning of variables
80 '
90  CLEAR:STATU$="i":RAD=3.141593/180: F=1:M=1:D$="b:"
100 DIM A(39,50),WW(50),WH(50),ACC(10),ACL(100),PB(50),PS(50),PHALP(50)
110 DIM ZW(50),FL(100),THETA(50),TV(50),TH(50),ZWT(50),SLOPE(50)
120 '
130 ' Definition of function keys
140 '
150 KEY OFF:FOR I = 1 TO 8:KEY I,"":NEXT I:KEY 1,"a":KEY 2,"b"
160 KEY 3,"c":KEY 4,"d":KEY 5,"e":KEY 6,"f":KEY 7,"g":KEY 8,"h"
170 '
180 ' Display of first page
190 '
200 SCREEN 0,0:WIDTH 80:CLS:LOCATE 5,17:COLOR 0,7:
210 PRINT " SARMA NON-VERTICAL SLICE STABILITY ANALYSIS "
220 COLOR 7,0:LOCATE 7,12
230 PRINT "Copyright - Evert Hoek, 1985.   This program is one of"
240 LOCATE 8,12
250 PRINT "a series of geotechnical programs developed as working"
260 LOCATE 9,12
270 PRINT "tools and for educational purposes. Use of the program"
280 LOCATE 10,12
290 PRINT "is not restricted  but the user is responsible for the"
300 LOCATE 11,12
310 PRINT "application of the results obtained from this program."
320 LOCATE 14,16:PRINT "Note: In order to operate this program a data"
330 LOCATE 15,16:PRINT "disk with at least one file with an extension"
340 LOCATE 16,16:PRINT ".SAR is required.   When starting a new disk,"
350 LOCATE 17,16:PRINT "ensure that such a file is stored on the disk"
360 LOCATE 18,16:PRINT "before it is used."
370 LOCATE 21,12
380 PRINT "Specify drive to be used for data disk (default B:)   "
390 LOCATE 21,65:INPUT " ",D$:IF LEN(D$)=0 THEN D$="b:"
400 IF RIGHT$(D$,1)=":" THEN D$=D$ ELSE D$=D$+":"
410 '
420 ' Display of second page
430 '
440 CLS:LOCATE 25,12:PRINT "to terminate input enter [q]";
450 LOCATE 25,41:PRINT "in response to any question":LOCATE 10,12
460 INPUT "Do you wish to read data from a disk file (y/n) ? :  ",DISK$
470 IF LEFT$(DISK$,1)="q" OR LEFT$(DISK$,1)="Q" THEN 6990
480 IF LEN(DISK$)=0 THEN 450
490 IF LEFT$(DISK$,1)="Y" OR LEFT$(DISK$,1)="y" THEN 680
500 LOCATE 11,12
510 INPUT "Number of slices to be included in analysis :       ",NUM$
520 IF LEFT$(NUM$,1)="q" OR LEFT$(NUM$,1)="Q" THEN 6990
530 IF LEN(NUM$)=0 THEN 500 ELSE NUM=VAL(NUM$)
540 FLAG2=0:LOCATE 12,12
550 INPUT "Unit weight of water =                         ",WATER$
560 IF LEFT$(WATER$,1)="q" OR LEFT$(WATER$,1)="Q" THEN 6990
570 IF LEN(WATER$)=0 THEN 540 ELSE WATER =VAL(WATER$)
580 IF WATER = 0 THEN FLAG2=1
590 FLAG3=0:LOCATE 13,12
600 INPUT "Are shear strengths uniform throughout slope (y/n) ? ",STRENGTH$
610 IF LEFT$(STRENGTH$,1)="q" OR LEFT$(STRENGTH$,1)="Q" THEN 6990
620 IF LEN(STRENGTH$)=0 THEN 590
```

```
630 IF LEFT$(STRENGTH$,1)="Y" OR LEFT$(STRENGTH$,1)="y" THEN FLAG3=1
640 N=NUM+1:FLAG6=0:GOTO 1860
650 '
660 ' Data entry from a disk file
670 '
680 CLS:LOCATE 3,1
690 PRINT STRING$(80,45):PRINT:PRINT "Sarma files on data disk :";
700 PRINT:FILES D$+"*.SAR":PRINT:PRINT STRING$(80,45):PRINT
710 INPUT "Enter filename (without extension): ",FILE$
720 OPEN D$+FILE$+".SAR" FOR INPUT AS #1
730 LINE INPUT#1, TITLE$:INPUT#1,N
740 INPUT#1,WATER:RAD=3.141593/180:F=1:M=1:NUM=N-1
750 INPUT#1,FLAG2:INPUT#1,FLAG3:INPUT#1,FLAG4
760 FOR K = 1 TO N:FOR J=1 TO 39:INPUT#1,A(J,K):NEXT J:NEXT K
770 CLOSE #1:FLAG6=1:STATUS$="e":FLAG50=1:CLS:GOTO 1860
780 '
790 ' Display of data array
800 '
810 CLS:LOCATE 1,1:PRINT "Analysis no. ";TITLE$
820 LOCATE 3,1:COLOR 15,0:PRINT "Side number":COLOR 7,0
830 LOCATE 4,1:PRINT "coordinate xt":LOCATE 5,1:PRINT "coordinate yt"
840 LOCATE 6,1:PRINT "coordinate xw":LOCATE 7,1:PRINT "coordinate yw"
850 LOCATE 8,1:PRINT "coordinate xb":LOCATE 9,1:PRINT "coordinate yb"
860 LOCATE 10,1:PRINT "friction angle":LOCATE 11,1:PRINT "cohesion"
870 LOCATE 12,1:PRINT "unit weight of water = ":LOCATE 12,23:PRINT WATER
880 LOCATE 13,1:COLOR 15,0:PRINT "Slice number":COLOR 7,0
890 LOCATE 14,1:PRINT "rock unit weight":LOCATE 15,1:PRINT "friction angle"
900 LOCATE 16,1:PRINT "cohesion":LOCATE 17,1:PRINT "force T"
910 LOCATE 18,1:PRINT "angle theta":GOSUB 960
920 IF STATUS$="i" THEN GOSUB 1060:RETURN ELSE RETURN
930 '
940 ' Subroutine for slice number display
950 '
960 COLOR 15,0:LOCATE 3,22:PRINT M:LOCATE 13,27:PRINT M
970 LOCATE 3,32:PRINT M+1:COLOR 7,0:IF N=M+1 THEN RETURN
980 COLOR 15,0:LOCATE 13,37:PRINT M+1:LOCATE 3,42
990 PRINT M+2:COLOR 7,0:IF N=M+2 THEN RETURN
1000 COLOR 15,0:LOCATE 13,47:PRINT M+2:LOCATE 3,52
1010 PRINT M+3:COLOR 7,0:IF N=M+3 THEN RETURN
1020 COLOR 15,0:LOCATE 13,57:PRINT M+3:LOCATE 3,62
1030 PRINT M+4:COLOR 7,0:IF N=M+4 THEN RETURN
1040 COLOR 15,0:LOCATE 13,67:PRINT M+4:LOCATE 3,72
1050 PRINT M+5:COLOR 7,0:RETURN
1060 LOCATE 20,13
1070 PRINT "Note: coordinates must increase from slope toe to crest"
1080 LOCATE 22,13
1090 PRINT "To edit title or data array, use direction keys to move"
1100 LOCATE 23,13
1110 PRINT "highlighted window.    Factor of safety calculation will"
1120 LOCATE 24,13
1130 PRINT "commence automatically when all data has been entered.";
1140 RETURN
1150 '
1160 ' Entry and display of title
1170 '
1180 IF FLAG6=1 OR STATUS$="e" THEN 1220
1190 LOCATE 1,1:COLOR 0,7:PRINT "Analysis no.":COLOR 7,0
1200 LOCATE 1,14:LINE INPUT "",TITLE$
1210 IF LEN(TITLE$)=0 THEN TITLE$=PREVT$
1220 LOCATE 1,1:PRINT "Analysis no. "
1230 IF STATUS$="r" THEN LOCATE 1,14:PRINT STRING$(66," ")
1240 LOCATE 1,14:PRINT TITLE$:PREVT$=TITLE$:RETURN
```

```
1250 '
1260 ' Subroutine for entry of data into array a(j,k)
1270 '
1280 IF STATU$="i" THEN FLAG6=0
1290 IF FLAG20=1 THEN FLAG6=1:FLAG20=0                      '1st column
1300 IF FLAG21=1 THEN K=M:J=PREVJ-1:FLAG21=0               'locate cursor
1310 IF FLAG23=1 THEN K=M+4:J=PREVJ-1:FLAG23=0             'locate cursor
1320 IF FLAG2=0 THEN 1350                                   'water present
1330 IF J=3 AND FLAG2=1 THEN J=5                            'drained
1340 IF FLAG2=1 THEN A(3,K)=A(5,K):A(4,K)=A(6,K)           'yw=yb
1350 IF J=7 AND K=1 THEN J=10:GOTO 1410                     'first slice
1360 IF FLAG3=0 THEN 1410                                   'variable strength
1370 IF J=7 THEN A(7,K)=A(11,1):FLAG6=1                     'uniform strength
1380 IF J=8 THEN A(8,K)=A(12,1):FLAG6=1
1390 IF J=9 AND FLAG12=0 THEN J=10                          'skip space
1400 IF J=10 THEN A(10,K)=A(10,1):FLAG6=1
1410 IF J=9 AND FLAG12=0 THEN J=10                          'skip space
1420 IF K=1 OR FLAG3=0 THEN 1450                            'variable strength
1430 IF J=11 THEN A(11,K)=A(11,1):FLAG6=1                   'uniform strength
1440 IF J=12 THEN A(12,K)=A(12,1):FLAG6=1
1450 IF J=13 THEN J=15                                      'skip space
1460 IF J=16 AND A(15,K)=0 THEN J=0:K=K+1:RETURN            'no forces
1470 IF J=17 THEN J=0:K=K+1:RETURN                          'next column
1480 IF J<=8 THEN X=18+(10*(K-M)):Y=J+3                     'cursor location
1490 IF J>=10 AND J<=12 THEN X=23+(10*(K-M)):Y=J+4
1500 IF J>=15 AND J<=16 THEN X=23+(10*(K-M)):Y=J+2
1510 IF FLAG6=1 THEN GOSUB 1790 ELSE GOSUB 1560            'cursor operation
1520 RETURN
1530 '
1540 ' Subroutine for cursor movement and display of array a(j,k)
1550 '
1560 FLAG10=0:FLAG11=0:FLAG12=0:FLAG13=0:FLAG14=0
1570 IF FLAG50=1 THEN FLAG50=0:F=1:GOTO 2960
1580 GOSUB 1810:Q$=INKEY$:IF Q$=" " THEN 1580              'scan keyboard
1590 IF LEN(Q$)=2 THEN Q$=RIGHT$(Q$,1)
1600 IF Q$="K" THEN GOSUB 1780:FLAG10=1:RETURN             'left
1610 IF Q$="M" THEN GOSUB 1780:FLAG11=1:RETURN             'right
1620 IF Q$="H" THEN GOSUB 1780:FLAG12=1:RETURN             'up
1630 IF Q$="P" THEN GOSUB 1780:FLAG13=1:RETURN             'down
1640 IF Q$="a" THEN GOSUB 1780:FLAG14=1:RETURN             'print
1650 IF Q$="b" THEN GOSUB 1780:FLAG15=1:RETURN             'calculate
1660 IF Q$="c" THEN GOSUB 1780:FLAG16=1:RETURN             'f.o.s vs K
1670 IF Q$="d" THEN GOSUB 1780:FLAG25=1:RETURN             'drain
1680 IF Q$="e" THEN GOSUB 1780:FLAG17=1:RETURN             'file
1690 IF Q$="f" THEN FLAG26=1:RETURN                        'restart
1700 IF Q$="g" OR Q$="q" THEN FLAG18=1:RETURN              'quit
1710 IF Q$="h" THEN FLAG27=1:RETURN                        'view
1720 IF Q$="0" THEN 1800                                   'enter zero
1730 IF Q$="-" THEN 1800                                   'enter minus
1740 IF Q$="." THEN 1800                                   'enter period
1750 IF VAL(Q$)<1 OR VAL(Q$)>9 THEN 1580                   'enter number
1760 LOCATE Y,X:PRINT VAL(Q$)
1770 LOCATE Y,X+2:INPUT "",IN$:A(J,K)=VAL(Q$+IN$)
1780 LOCATE Y,X:PRINT "              "
1790 LOCATE Y,X:PRINT USING "#####.##";A(J,K):RETURN       'display entry
1800 LOCATE Y,X:PRINT " ";:PRINT Q$:GOTO 1770
1810 LOCATE Y,X:COLOR 0,7
1820 PRINT "          ":COLOR 7,0:RETURN
1830 '
1840 ' Data entry and display of array a(j,k)
1850 '
1860 F=1:M=1:GOSUB 810:GOSUB 1180                          'screen display
```

```
1870 FOR K=1 TO 6:FOR J=1 TO 17
1880 IF FLAG1=1 THEN 2050
1890 IF FLAG23=1 THEN FLAG6=0:GOTO 1910
1900 IF FLAG22=1 THEN FLAG6=1:GOTO 2050
1910 IF FLAG10=1 THEN GOSUB 2420                       'left
1920 IF FLAG11=1 THEN GOSUB 2530                       'right
1930 IF FLAG12=1 THEN GOSUB 2640                       'up
1940 IF FLAG13=1 THEN GOSUB 2760                       'down
1950 IF FLAG14=1 THEN GOSUB 5580                       'print
1960 IF FLAG15=1 THEN F=1:GOTO 2960                    'calculate
1970 IF FLAG16=1 THEN 5050                             'f.o.s vs K
1980 IF FLAG17=1 THEN 6160                             'file
1990 IF FLAG18=1 THEN 6990                             'quit
2000 IF FLAG19=1 THEN 2100                             'next page
2010 IF FLAG25=1 THEN FLAG25=0:GOSUB 2860             'drain
2020 IF FLAG26=1 THEN 2030 ELSE 2040
2030 FLAG26=0:STATUS$="i":TITLE$="":GOTO 440          'restart
2040 IF FLAG27=1 THEN 6280                             'view
2050 IF K<N THEN 2070                                  'check end
2060 IF J=7 THEN 2950                                  'end input
2070 IF K=6 AND J=9 THEN 2090                          'first page
2080 GOSUB 1280:NEXT J:NEXT K                          'cursor operation
2090 FLAG1=0:IF STATUS$="e" THEN 4160                  'display 1st page
2100 M=M+5:GOSUB 810                                   'renumber columns
2110 FOR K=M TO M+5: FOR J=1 TO 17                     'second page
2120 IF STATUS$="e" THEN 2140
2130 IF K=M AND J<=8 THEN FLAG20=1:GOTO 2380
2140 IF FLAG19=1 THEN FLAG6=1:GOTO 2340
2150 IF FLAG21=1 THEN FLAG6=0:GOTO 2380
2160 IF FLAG22=1 THEN FLAG6=1:GOTO 2340
2170 IF FLAG23=1 THEN FLAG6=0:GOTO 2380
2180 IF FLAG10=1 THEN GOSUB 2420                       'left
2190 IF FLAG11=1 THEN GOSUB 2530                       'right
2200 IF FLAG12=1 THEN GOSUB 2640                       'up
2210 IF FLAG13=1 THEN GOSUB 2760                       'down
2220 IF FLAG14=1 THEN GOSUB 5580                       'print
2230 IF FLAG15=1 THEN F=1:GOTO 2960                    'calculate
2240 IF FLAG16=1 THEN 5050                             'f.o.s vs K
2250 IF FLAG17=1 THEN 6160                             'file
2260 IF FLAG18=1 THEN 6990                             'quit
2270 IF FLAG25=1 THEN FLAG25=0:GOSUB 2860             'drain
2280 IF FLAG26=1 THEN 2290 ELSE 2300
2290 FLAG26=0:STATUS$="i":TITLE$="":GOTO 440          'restart
2300 IF FLAG27=1 THEN 6280                             'view
2310 IF FLAG19=1 THEN K=M:J=1:GOTO 2100
2320 IF FLAG22=1 AND M=6 THEN M=1:GOTO 1860
2330 IF FLAG22=1 AND M>=11 THEN K=M:J=1:M=M-10:GOTO 2100
2340 IF K<N THEN 2360                                  'check end
2350 IF J=7 THEN 2950                                  'end input
2360 IF K=M+5 AND J=9 THEN 2370 ELSE 2380
2370 IF STATUS$="i" THEN 2100 ELSE 2950
2380 GOSUB 1280:NEXT J:NEXT K
2390 '
2400 ' Subroutine to move cursor left
2410 '
2420 IF K=2 AND J=8 THEN K=K:GOTO 2480                 'limit left
2430 IF K=2 AND J=9 THEN K=K:GOTO 2480                 'limit left
2440 IF K=1 THEN K=1:GOTO 2480
2450 IF M>=6 AND K=M THEN 2490                         'previous page
2460 IF K>M THEN K=K-1                                 'left
2470 IF J=17 AND A(16,K)=0 THEN K=K+1:GOTO 2480        'blank cell
2480 J=J-1:FLAG10=0:RETURN                             'keep line
```

```
2490 FLAG22=1:PREVJ=J:FLAG10=0:RETURN
2500 '
2510 ' Subroutine to move cursor right
2520 '
2530 IF K=M+5 AND J<=9 THEN 2600                          'next page
2540 IF K=M+4 AND J>9 THEN 2600                           'next page
2550 IF K<N THEN K=K+1                                    'right
2560 IF J=17 AND A(16,K)=0 THEN K=K-1:GOTO 2590          'blank cell
2570 IF J<=7 AND K=N THEN K=N                             'limit right
2580 IF J>7 AND K=N THEN K=NUM                            'limit right
2590 J=J-1:FLAG11=0:RETURN                                'keep line
2600 FLAG19=1:PREVJ=J:FLAG11=0:RETURN
2610 '
2620 ' Subroutine to move cursor up
2630 '
2640 IF J=2 AND K=1 THEN 2650 ELSE 2660
2650 STATU$="r":GOSUB 1190:STATU$="e":J=1:K=1:GOTO 2720  'edit title
2660 IF J=2 THEN J=1:GOTO 2720                            'limit up
2670 IF J=6 AND FLAG2=1 THEN J=2:GOTO 2720               'drained
2680 IF K=1 AND J=11 THEN J=6:GOTO 2720                  '1st column
2690 IF J=11 THEN J=8:GOTO 2720                           'skip space
2700 IF J=16 THEN J=12:GOTO 2720                          'skip space
2710 IF J>=3 AND J<=17 THEN J=J-2:FLAG12=0:GOTO 2720     'up
2720 FLAG12=0:RETURN
2730 '
2740 ' Subroutine to move cursor down
2750 '
2760 IF K=N AND J=7 THEN J=6:GOTO 2820                   'last column
2770 IF FLAG2=1 AND J=3 THEN J=5:GOTO 2820               'skip space
2780 IF J=9 THEN J=10:GOTO 2820                           'skip space
2790 IF J=13 THEN J=15:GOTO 2820                          'skip space
2800 IF J=16 AND A(J,K)=0 THEN J=15:GOTO 2820            'down limit
2810 IF J=17 THEN J=16:GOTO 2820                          'down limit
2820 FLAG13=0:RETURN
2830 '
2840 ' Drain by changing unit weight of water
2850 '
2860 LOCATE 12,1:PRINT STRING$(30," "):COLOR 0,7
2870 LOCATE 12,1:PRINT "unit weight of water"
2880 COLOR 7,0:LOCATE 12,23:INPUT "= ",WATER
2890 LOCATE 12,1:PRINT STRING$(30," "):LOCATE 12,1
2900 PRINT "unit weight of water = ";WATER
2910 J=J-1:F=1:GOTO 2960
2920 '
2930 ' Calculation of slice parameters
2940 '
2950 IF STATU$="e"  THEN 4160
2960 GOSUB 3220:FOR K=1 TO N:GOSUB 3330:NEXT K           'd & delta
2970 FOR K=1 TO NUM:GOSUB 3390:NEXT K                    'b,alpha,W & U
2980 WAT=.5*WATER:GOSUB 3480                             'water forces
2990 FOR K=1 TO N:A(24,K)=A(12,K)/F:A(26,K)=A(8,K)/F    'cb/F,cs/F
3000 PB(K)=A(11,K)*RAD:PS(K)=A(7,K)*RAD                  'deg to radians
3010 A(25,K)=TAN(PB(K))/F                                'tanphi/F
3020 A(27,K)=TAN(PS(K))/F                                'tanphi/F
3030 PB(K)=ATN(A(25,K)):PS(K)=ATN(A(27,K))              'effective phi
3040 PHALP(K)=PB(K)-A(20,K):THETA(K)=A(16,K)*RAD        'phi-alpha
3050 TV(K)=A(15,K)*SIN(THETA(K))+WW(K)                   'TV
3060 IF A(16,K)=90 THEN TH(K)=0:GOTO 3090               'vertical
3070 IF A(16,K)=270 THEN TH(K)=0:GOTO 3090              'force
3080 TH(K)=A(15,K)*COS(THETA(K))                         'TH
3090 TH(K)=TH(K)+WH(K):NEXT K
3100 TV(N-1)=TV(N-1)-A(23,N)*SIN(A(18,N))               'water in
```

```
3110 TH(N-1)=TH(N-1)-A(23,N)*COS(A(18,N))          'tension crack
3120 TV(1)=TV(1)+A(23,1)*SIN(A(18,1))              'submerged toe
3130 TH(1)=TH(1)+A(23,1)*COS(A(18,1))              'submerged toe
3140 '
3150 '  Calculation of Kc
3160 '
3170 FOR K=2 TO N:GOSUB 3780:NEXT K                'S
3180 FOR K=1 TO NUM:GOSUB 3790:NEXT K:GOTO 3950    'R,Q,e,P,a.Kc
3190 '
3200 '  Subroutine for display of "calculating"
3210 '
3220 IF FLAG15=0 THEN LOCATE 20,7:PRINT STRING$(70," ")
3230 LOCATE 22,7:PRINT STRING$(70," ")
3240 LOCATE 23,7:PRINT STRING$(70," ")
3250 LOCATE 24,7:PRINT STRING$(70," ");
3260 LOCATE 22,28:COLOR 0,7
3270 PRINT " C A L C U L A T I N G ":COLOR 7,0
3280 FOR K=1 TO N:TV(K)=0:TH(K)=0:WW(K)=0:WH(K)=0
3290 ZW(K)=0:ZWT(K)=0:NEXT K:RETURN
3300 '
3310 '  Subroutines for calculation of slice geometry
3320 '
3330 IF A(4,K)<A(6,K) THEN A(4,K)=A(6,K):A(3,K)=A(5,K)  'check water
3340 DSQ=(A(1,K)-A(5,K))^2+(A(2,K)-A(6,K))^2
3350 IF DSQ=0 THEN A(17,K)=0 ELSE A(17,K)=SQR(DSQ)      'd
3360 IF A(2,K)-A(6,K)=0 THEN A(18,K)=0:RETURN
3370 A(18,K)=ATN((A(1,K)-A(5,K))/(A(2,K)-A(6,K)))       'delta
3380 RETURN
3390 A(19,K)=A(5,K+1)-A(5,K)                            'b
3400 IF A(19,K)=0 THEN A(20,K)=0:GOTO 3420
3410 A(20,K)=ATN((A(6,K+1)-A(6,K))/A(19,K))             'alpha
3420 A(21,K)=(A(6,K)-A(2,K+1))*(A(1,K)-A(5,K+1))
3430 A(21,K)=A(21,K)+(A(2,K)-A(6,K+1))*(A(1,K+1)-A(5,K))
3440 A(21,K)=.5*A(10,K)*A(21,K):RETURN                  'W
3450 '
3460 '  Subroutine for calculation of water forces
3470 '
3480 FOR K=1 TO NUM:ZW(K)=A(4,K)-A(6,K)
3490 ZW(K+1)=A(4,K+1)-A(6,K+1)
3500 A(22,K)=WAT*(ZW(K)+ZW(K+1))*A(19,K)
3510 A(22,K)=ABS(A(22,K)/COS(A(20,K))):NEXT K           'U
3520 FOR K=1 TO N:ZWT(K)=A(4,K)-A(2,K)
3530 IF ZWT(K)>0 THEN 3550
3540 A(23,K)=WAT*ABS(ZW(K)^2/COS(A(18,K))):GOTO 3570    'PW
3550 A(23,K)=WAT*(ZWT(K)+ZW(K))                         'submerged
3560 A(23,K)=A(23,K)*ABS((A(2,K)-A(6,K))/COS(A(18,K)))  'PW
3570 NEXT K
3580 FOR K=1 TO NUM
3590 IF ZWT(K)>=0 AND ZWT(K+1)>=0 THEN 3600 ELSE 3640
3600 WW(K)=WAT*(ZWT(K)+ZWT(K+1))                        'WW fully
3610 WW(K)=WW(K)*ABS((A(1,K+1)-A(1,K)))                 'submerged
3620 WH(K)=WAT*(A(2,K+1)-A(2,K))*(ZWT(K)+ZWT(K+1))      'WH
3630 IF A(2,K+1)<A(2,K) THEN WH(K)=-WH(K):GOTO 3740
3640 IF ZWT(K)>=0 AND ZWT(K+1)<=0 THEN 3650 ELSE 3690
3650 WW(K)=WAT*ZWT(K)^2*(A(1,K+1)-A(1,K))               'WW side i
3660 IF A(2,K+1)-A(2,K)=0 THEN WW(K)=0:GOTO 3680
3670 WW(K)=ABS(WW(K)/(A(2,K+1)-A(2,K)))                 'submerged
3680 WH(K)=WAT*ZWT(K)^2:GOTO 3740                       'WH
3690 IF ZWT(K)<=0 AND ZWT(K+1)>=0 THEN 3700 ELSE 3740
3700 WW(K)=WAT*ZWT(K+1)^2*(A(1,K+1)-A(1,K))             'WW side i+1
3710 IF A(2,K+1)-A(2,K)=0 THEN WW(K)=0:GOTO 3730
3720 WW(K)=ABS(WW(K)/(A(2,K+1)-A(2,K)))                 'submerged
```

```
3730 WH(K)=-WAT*(ZWT(K+1))^2
3740 NEXT K:RETURN
3750 '
3760 ' Subroutines for calculation of S,R,Q,e,P and a
3770 '
3780 A(29,K)=A(26,K)*A(17,K)-A(23,K)*A(27,K):RETURN          ' S
3790 A(28,K)=A(24,K)*A(19,K)
3800 A(28,K)=A(28,K)/COS(A(20,K))-A(22,K)*A(25,K)            ' R
3810 A(30,K)=COS(PB(K)-A(20,K)+PS(K+1)-A(18,K+1))
3820 A(30,K)=COS(PS(K+1))/A(30,K)                            ' Q
3830 A(31,K)=A(30,K)*COS(PB(K)-A(20,K)+PS(K)-A(18,K))
3840 A(31,K)=A(31,K)/COS(PS(K))                              ' e
3850 A(32,K)=A(30,K)*A(21,K)*COS(PB(K)-A(20,K))              ' P
3860 A(33,K)=(A(21,K)+TV(K))*SIN(PHALP(K))
3870 A(33,K)=A(33,K)+TH(K)*COS(PHALP(K))
3880 A(33,K)=A(33,K)+A(28,K)*COS(PB(K))
3890 A(33,K)=A(33,K)+A(29,K+1)*SIN(PHALP(K)-A(18,K+1))
3900 A(33,K)=A(33,K)-A(29,K)*SIN(PHALP(K)-A(18,K))
3910 A(33,K)=A(33,K)*A(30,K):RETURN                          ' a
3920 '
3930 ' Calculation of Kc and FOS
3940 '
3950 GOSUB 4470
3960 IF FLAG7=1 OR FLAG8=1 OR FLAG77=1 THEN 3990
3970 IF F<>1 THEN 4080
3980 IF (Z2+A(32,NUM))=0 THEN ACC=0:GOTO 4110
3990 ACC(1)=(Z3+A(33,NUM))/(Z2+A(32,NUM))                    ' Kc
4000 IF F=1 THEN ACC=ACC(1)
4010 IF FLAG8=1 THEN 4110
4020 IF FLAG7 = 1 THEN 5210
4030 IF FLAG77 = 1 THEN 6890
4040 F=1+3.33*ACC(1)                                         ' FOS estimate
4050 IF F<=0 THEN F=.1
4060 IF F>5 THEN F=5
4070 GOTO 2990                                               ' recalculate K
4080 ACC(2)=(Z3+A(33,NUM))/(Z2+A(32,NUM))                    ' new K
4090 Y=1/F:FS=1-ACC(1)*(1-Y)/(ACC(1)-ACC(2)):FOS=1/FS        ' FOS
4100 F=FOS:FLAG8=1:GOTO 2990
4110 FLAG8=0:FLAG4=0:FOR K=1 TO NUM:GOSUB 4740:NEXT K        ' normal stresses
4120 FOR K=1 TO NUM:GOSUB 4770:NEXT K
4130 IF M>=6 THEN 4140 ELSE 4150
4140 F=1:M=1:STATUS$="e":FLAG6=1:FLAG15=0:GOTO 1860
4150 LOCATE 22,28:PRINT "                               "
4160 IF FLAG50=1 THEN 1570
4170 GOSUB 4950:LOCATE 22,7:COLOR 15,0
4180 LOCATE 22,7:COLOR 15,0
4190 PRINT "Acceleration Kc  = ";
4200 LOCATE 22,27:PRINT USING "##.####";ACC;
4210 LOCATE 22,47:PRINT "Factor of safety = ";
4220 LOCATE 22,66:PRINT USING "##.##";FOS;:COLOR 7,0
4230 IF FLAG4=0 AND ABS(ACC(2))>.1 THEN 4240 ELSE 4270
4240 LOCATE 23,7:COLOR 15
4250 PRINT "Large extrapolation  -   plot of fos vs K suggested";
4260 LOCATE 23,59:PRINT "to check fos";:COLOR 7,0
4270 IF FLAG4=0 THEN 4340 ELSE LOCATE 23,7:COLOR 15,0
4280 PRINT "Negative effective normal stresses    - "
4290 LOCATE 23,50:PRINT "solution unacceptable"
4300 COLOR 7,0
4310 '
4320 ' Control of screen displays
4330 '
4340 STATUS$="e":FLAG15=0:GOSUB 4540
```

```
4350 IF STATU$="e" THEN 4370
4360 FLAG6=1:STATU$="e":GOTO 1860
4370 IF FLAG19=1 THEN 4380 ELSE 4390              'next page
4380 FLAG21=1:FLAG3=0:FLAG19=0:GOTO 2110
4390 IF FLAG22=1 AND K=7 THEN 4400 ELSE 4410      'previous page
4400 FLAG23=1:FLAG3=0:FLAG22=0:GOTO 1870
4410 IF FLAG22=1 AND K>=11 THEN 4420 ELSE 4430    'previous page
4420 FLAG23=1:FLAG3=0:FLAG22=0:GOTO 2110
4430 FLAG6=0:FLAG3=0:J=1:K=1:M=1:GOTO 1870
4440 '
4450 ' Subroutine for calculation of K
4460 '
4470 Z1=1:Z2=0:Z3=0
4480 FOR K=NUM TO 2 STEP-1
4490 Z1=Z1*A(31,K):Z2=Z2+A(32,K-1)*Z1
4500 Z3=Z3+A(33,K-1)*Z1:NEXT K:RETURN
4510 '
4520 ' Display of function key
4530 '
4540 LOCATE 25,1:PRINT "1";
4550 LOCATE 25,3:COLOR 0,7:PRINT "print";
4560 COLOR 7,0:LOCATE 25,10:PRINT "2";
4570 LOCATE 25,12:COLOR 0,7:PRINT "calculate";
4580 COLOR 7,0:LOCATE 25,23:PRINT "3";
4590 LOCATE 25,25:COLOR 0,7:PRINT "fos vs K";
4600 COLOR 7,0:LOCATE 25,35:PRINT "4";
4610 LOCATE 25,37:COLOR 0,7:PRINT "drain";
4620 COLOR 7,0:LOCATE 25,44:PRINT "5";
4630 LOCATE 25,46:COLOR 0,7:PRINT "file";
4640 COLOR 7,0:LOCATE 25,52:PRINT "6";
4650 LOCATE 25,54:COLOR 0,7:PRINT "restart";
4660 COLOR 7,0:LOCATE 25,63:PRINT "7";
4670 LOCATE 25,65:COLOR 0,7:PRINT "quit";
4680 COLOR 7,0:LOCATE 25,70:PRINT "8";
4690 LOCATE 25,72:COLOR 0,7:PRINT "view";
4700 COLOR 7,0:RETURN
4710 '
4720 ' Subroutine for calculation of effective normal stresses
4730 '
4740 A(34,K+1)=A(33,K)+A(34,K)*A(31,K)-ACC(1)*A(32,K)      'Ei+1
4750 A(35,K)=(A(34,K)-A(23,K))*A(27,K)+A(26,K)*A(17,K)     'Xi
4760 RETURN
4770 A(36,K)=A(21,K)+TV(K)+A(35,K+1)*COS(A(18,K+1))
4780 A(36,K)=A(36,K)-A(35,K)*COS(A(18,K))
4790 A(36,K)=A(36,K)-A(34,K+1)*SIN(A(18,K+1))
4800 A(36,K)=A(36,K)+A(34,K)*SIN(A(18,K))
4810 A(36,K)=A(36,K)+A(22,K)*A(25,K)*SIN(A(20,K))
4820 A(36,K)=A(36,K)-A(24,K)*A(19,K)*TAN(A(20,K))
4830 A(36,K)=A(36,K)*COS(PB(K))/COS(PHALP(K))              'Ni
4840 A(37,K)=(A(36,K)-A(22,K))*A(25,K)
4850 A(37,K)=A(37,K)+A(24,K)*A(19,K)/COS(A(20,K))          'TSi
4860 A(38,K)=(A(36,K)-A(22,K))*COS(A(20,K))/A(19,K)        'sigma b
4870 IF A(17,K)=0 THEN A(39,K)=0:GOTO 4900
4880 IF K=1 THEN A(39,K)=0:GOTO 4900
4890 A(39,K)=(A(34,K)-A(23,K))/A(17,K)                     'sigma s
4900 IF A(38,K)<0 OR A(39,K)<0 THEN FLAG4=1
4910 RETURN
4920 '
4930 ' Subroutine for display of normal stresses
4940 '
4950 LOCATE 19,1:PRINT "base stresses"
4960 LOCATE 20,1:PRINT "side stresses"
```

```
4970 MEND=M+5:IF MEND<N THEN 4980 ELSE MEND=N
4980 J=38:FOR K=M TO MEND-1:X=23+(10*(K-M)):Y=19
4990 LOCATE Y,X:PRINT USING "#####.##";A(J,K):NEXT K
5000 J=39:FOR K=M TO MEND:X=18+(10*(K-M)):Y=20
5010 LOCATE Y,X:PRINT USING "#####.##";A(J,K):NEXT K:RETURN
5020 '
5030 ' Calculation of acceleration K for different safety factors
5040 '
5050 CLS:SCREEN 2:GOSUB 6730
5060 LOCATE 3,57:PRINT "factor of safety"
5070 LOCATE 4,53:PRINT " versus acceleration K"
5080 LOCATE 8,54:PRINT " PLEASE WAIT FOR PLOT":GOSUB 6870
5090 LOCATE 8,54:PRINT STRING$(21," ")
5100 LOCATE 25,2:PRINT "to terminate calculation press ";
5110 PRINT "[ENTER]";
5120 PRINT " in response to prompt for a new value";
5130 LOCATE 6,54:PRINT "Enter f.o.s. = ";
5140 LOCATE 8,57:PRINT "f.o.s";:LOCATE 8,68:PRINT "acc. K";
5150 Y=6:X=69:LOCATE Y,X:INPUT " ",F$
5160 IF LEN(F$)=0 THEN 5260
5170 IF F$="0" THEN F$=".01"
5180 F=VAL(F$):FL(L)=F:Y=10:X=56
5190 LOCATE Y,X:PRINT USING "##.####";FL(L)
5200 FLAG7 = 1:GOTO 2990
5210 ACL(L)=ACC(1):L=L+1
5220 IF FLAG30=1 THEN 5410
5230 LOCATE Y,67:PRINT USING "##.####";ACC(1)
5240 LOCATE 6,69:PRINT "          ";
5250 FLAG7 = 0:GOTO 5150
5260 FIN=L:LOCATE 25,1:PRINT STRING$(78," ");
5270 LOCATE 25,1:PRINT "[F1]";:LOCATE 25,6
5280 PRINT "print fos vs K";
5290 LOCATE 25,23:PRINT "[F2]";:LOCATE 25,28
5300 PRINT "return to slice data array";
5310 LOCATE 25,57:PRINT "[F3]";:LOCATE 25,62
5320 PRINT "restart";
5330 LOCATE 25,71:PRINT "[F4]";:LOCATE 25,76
5340 PRINT "quit";:FLAG16=0
5350 Q$=INKEY$:IF Q$=" " THEN 5350
5360 IF Q$ = "a" THEN GOSUB 5450:GOTO 5350
5370 IF Q$ = "b" THEN SCREEN 0,0,0:CHECK=1:GOTO 5400
5380 IF Q$ = "c" THEN SCREEN 0,0,0:GOTO 440
5390 IF Q$ = "d" THEN SCREEN 0,0,0:GOTO 6990 ELSE GOTO 5350
5400 F=1:FLAG30=1:GOTO 5200
5410 FLAG30=0:FLAG7=0:STATU$="e":FLAG6=1:F=1:M=1:GOTO 1860
5420 '
5430 ' Subroutine for printing fos vs K
5440 '
5450 LPRINT:LPRINT:LPRINT:LPRINT
5460 LPRINT TAB(13) "Analysis no. ";:LPRINT TITLE$
5470 LPRINT TAB(13) "Plot of factor of safety";
5480 LPRINT TAB(38) "versus acceleration K"
5490 LPRINT:LPRINT TAB(19) "f.o.s";
5500 LPRINT TAB(32) "acc. K";:LPRINT TAB(44) "1/fos":PRINT
5510 FOR L=1 TO FIN-1:LPRINT TAB(18) USING "##.####";FL(L);
5520 LPRINT TAB(31) USING "##.####";ACL(L);
5530 LPRINT TAB(43) USING "##.####";1/FL(L):NEXT L
5540 LPRINT:LPRINT:LPRINT:LPRINT:RETURN
5550 '
5560 ' Subroutine for printing array and results
5570 '
5580 PREVJ=J:PREVK=K:PREVM=M:FLAG14=0
```

```
5590 LPRINT:LPRINT:LPRINT:LPRINT:LPRINT:LPRINT
5600 LPRINT TAB(23) "SARMA NON-VERTICAL SLICE ANALYSIS":LPRINT
5610 LPRINT "Analysis no. ";:LPRINT TAB(14) TITLE$
5620 LPRINT:LPRINT "Unit weight of water =";
5630 LPRINT TAB(23) WATER:X1=1:X2=N
5640 IF X2>6 THEN X2=X1+5
5650 T1=20:LPRINT:LPRINT "Side number";
5660 FOR X=X1 TO X2:LPRINT TAB(T1);:LPRINT USING "##";X;
5670 T1=T1+10:NEXT X:X3=X2:IF X2=N THEN LPRINT
5680 T1=16:LPRINT "Coordinate xt";:J=1:GOSUB 6070
5690 T1=16:LPRINT "Coordinate yt";:J=2:GOSUB 6070
5700 T1=16:LPRINT "Coordinate xw";:J=3:GOSUB 6070
5710 T1=16:LPRINT "Coordinate yw";:J=4:GOSUB 6070
5720 T1=16:LPRINT "Coordinate xb";:J=5:GOSUB 6070
5730 T1=16:LPRINT "Coordinate yb";:J=6:GOSUB 6070
5740 T1=16:LPRINT "Friction angle";:J=7:GOSUB 6070
5750 T1=16:LPRINT "Cohesion";:J=8:GOSUB 6070
5760 T1=25:LPRINT:LPRINT "Slice number";
5770 IF X2=N THEN X3=N-1 ELSE X3=X2
5780 FOR X=X1 TO X3:LPRINT TAB(T1);:LPRINT USING "##";X;
5790 T1=T1+10:NEXT X:IF X2=N THEN LPRINT
5800 T1=21:LPRINT "Rock unit weight";:J=10:GOSUB 6070
5810 T1=21:LPRINT "Friction angle";:J=11:GOSUB 6070
5820 T1=21:LPRINT "Cohesion";:J=12:GOSUB 6070
5830 T1=21:LPRINT "Force T";:J=15:GOSUB 6070
5840 T1=21:LPRINT "Angle theta";:J=16:GOSUB 6070:LPRINT
5850 LPRINT "Effective normal stresses "
5860 T1=21:LPRINT "Base";:J=38:GOSUB 6070
5870 T1=16:LPRINT "Side";:J=39:GOSUB 6070:LPRINT:LPRINT
5880 IF X2=N THEN 5950
5890 IF X2<N THEN X1=X1+6:X2=N
5900 IF X2<X1+5 THEN 5910 ELSE X2=X1+5
5910 IF X1=13 OR X1=27 THEN 5930
5920 LPRINT:LPRINT:GOTO 5650
5930 LPRINT:LPRINT:LPRINT:LPRINT:LPRINT
5940 LPRINT:LPRINT:LPRINT:LPRINT:GOTO 5650
5950 LPRINT TAB(7) "Acceleration Kc = ";
5960 LPRINT TAB(27);:LPRINT USING "##.####";ACC;
5970 LPRINT TAB(47) "Factor of Safety = ";
5980 LPRINT TAB(66);:LPRINT USING "##.##";FOS
5990 IF FLAG4=0 THEN 6020
6000 LPRINT TAB(7) "Negative effective normal stresses";
6010 LPRINT TAB(45) "-   solution unacceptable":GOTO 6050
6020 IF FLAG4=0 AND ABS(ACC(2))<.1 THEN 6050
6030 LPRINT TAB(7) "Large extrapolation - plot of fos";
6040 LPRINT TAB(44) "vs K suggested to check fos"
6050 J=PREVJ-1:K=PREVK:M=PREVM
6060 LPRINT:LPRINT:LPRINT:RETURN
6070 FOR K=X1 TO X3:LPRINT TAB(T1);:GOSUB 6090
6080 T1=T1+10:NEXT K:LPRINT:RETURN
6090 IF A(J,K)=0 THEN 6110
6100 IF ABS(A(J,K))>99999! OR ABS(A(J,K))<8.999999E-03 THEN 6120
6110 LPRINT USING "#####.##";A(J,K);:RETURN
6120 LPRINT USING "##.##^^^^";A(J,K);:RETURN
6130 '
6140 ' Storage of data on disk file
6150 '
6160 CLS:LOCATE 4,1:PRINT STRING$(80,45)
6170 PRINT:PRINT "Sarma files on data disk :";
6180 PRINT:FILES D$+"*.SAR":PRINT:PRINT STRING$(80,45):PRINT
6190 INPUT "Enter filename (without extension): ",FILE$
6200 OPEN D$+FILE$+".SAR" FOR OUTPUT AS #2
```

```
6210 PRINT#2,TITLE$:PRINT#2,N:WRITE#2,WATER
6220 WRITE#2,FLAG2:WRITE#2,FLAG3:WRITE#2,FLAG4
6230 FOR K=1 TO N:FOR J=1 TO 39:WRITE#2,A(J,K):NEXT J:NEXT K
6240 CLOSE#2:FLAG6=1:STATU$="e":F=1:M=1:FLAG17=0:GOTO 1860
6250 '
6260 ' Graphical display of geometry
6270 '
6280 SCREEN 2:XA=0:XB=0:YA=0:YB=0:JMIN=1:JMAX=5:DJ=2
6290 KMIN=1:KMAX=N:DK=N-1:GOSUB 6630:XMIN=MIN                    'min x
6300 JMIN=1:JMAX=5:DJ=2:KMIN=1:KMAX=N
6310 DK=N-1:GOSUB 6690:XMAX=MAX                                  'max x
6320 JMIN=2:JMAX=6:DJ=2:KMIN=1:KMAX=N
6330 DK=1:GOSUB 6630:YMIN=MIN                                    'min y
6340 JMIN=2:JMAX=6:DJ=2:KMIN=1:KMAX=N
6350 DK=1:GOSUB 6690:YMAX=MAX                                    'max y
6360 XSC=270/(XMAX-XMIN):YSC=160/(YMAX-YMIN)                     'scale factor
6370 IF XSC<YSC THEN SC=XSC ELSE SC=YSC
6380 LNY=199:XADJ=(319/SC-(XMAX-XMIN))/2                         'center plot
6390 YADJ=(199/SC-(YMAX-YMIN))/2
6400 XMIN=XMIN-XADJ:YMIN=YMIN-YADJ:FOR K=1 TO N-1
6410 XA=2*(A(1,K)-XMIN)*SC:YA=LNY-(A(2,K)-YMIN)*SC
6420 XB=2*(A(1,K+1)-XMIN)*SC:YB=LNY-(A(2,K+1)-YMIN)*SC
6430 LINE (XA,YA)-(XB,YB),3                                      'top surface
6440 XA=2*(A(5,K)-XMIN)*SC:YA=LNY-(A(6,K)-YMIN)*SC
6450 XB=2*(A(5,K+1)-XMIN)*SC:YB=LNY-(A(6,K+1)-YMIN)*SC
6460 LINE (XA,YA)-(XB,YB),3                                      'failure surface
6470 XA=2*(A(1,K)-XMIN)*SC:YA=LNY-(A(2,K)-YMIN)*SC
6480 XB=2*(A(5,K)-XMIN)*SC:YB=LNY-(A(6,K)-YMIN)*SC
6490 LINE (XA,YA)-(XB,YB),3                                      'slice sides
6500 XA=2*(A(3,K)-XMIN)*SC:YA=LNY-(A(4,K)-YMIN)*SC
6510 XB=2*(A(3,K+1)-XMIN)*SC:YB=LNY-(A(4,K+1)-YMIN)*SC
6520 LINE (XA,YA)-(XB,YB),1:NEXT K                               'water surface
6530 XA=2*(A(1,N)-XMIN)*SC:YA=LNY-(A(2,N)-YMIN)*SC
6540 XB=2*(A(5,N)-XMIN)*SC:YB=LNY-(A(6,N)-YMIN)*SC
6550 LINE (XA,YA)-(XB,YB),3                                      'nth slice side
6560 LOCATE 25,29:PRINT "press any key to return";
6570 IF LEN(INKEY$)=0 THEN 6570
6580 FLAG27=0:SCREEN 0,0,0:WIDTH 80                              'reset screen
6590 STATU$="e":FLAG6=1:F=1:M=1:GOTO 1860                        'return
6600 '
6610 ' Subroutine to find MIN in range
6620 '
6630 MIN=A(JMIN,KMIN):FOR J=JMIN TO JMAX STEP DJ
6640 FOR K=KMIN TO KMAX STEP DK:IF MIN>A(J,K) THEN MIN=A(J,K)
6650 NEXT K:NEXT J:RETURN
6660 '
6670 ' Subroutine to find MAX in range
6680 '
6690 MAX=A(JMIN,KMIN):FOR J=JMIN TO JMAX STEP DJ
6700 FOR K=KMIN TO KMAX STEP DK:IF MAX<A(J,K) THEN MAX=A(J,K)
6710 NEXT K:NEXT J:RETURN
6720 '
6730 'Subroutine to plot fos vs K
6740 '
6750 LINE (5,5)-(635,185),,B:LINE (395,8)-(630,90),,B
6760 LINE (200,5)-(200,185):LINE (5,125)-(635,125)
6770 LINE (66,125)-(66,130):LOCATE 17,7:PRINT "-0.5"
6780 LINE (333,125)-(333,130):LOCATE 17,41:PRINT "0.5"
6790 LINE (466,125)-(466,130):LOCATE 17,58:PRINT "1.0"
6800 LINE (600,125)-(600,130):LOCATE 17,75:PRINT "1.5"
6810 LINE (200,13)-(205,13):LOCATE 2,28:PRINT "2.5"
6820 LINE (200,50)-(205,50):LOCATE 7,28:PRINT "2.0"
```

```
6830 LINE (200,87)-(205,87):LOCATE 12,28:PRINT "1.5"
6840 LINE (200,166)-(205,166):LOCATE 22,28:PRINT "0.5"
6850 LOCATE 3,24:PRINT "f":LOCATE 4,24:PRINT "o":LOCATE 5,24
6860 PRINT "s":LOCATE 15,60:PRINT "acceleration K":FLAG77=0:RETURN
6870 F=2.5:L=1
6880 FL(L)=F:ACL(L)=ACC(1):FLAG77=1:GOTO 2990
6890 KSX=(ACC(1)*400/1.5)+200:FSY=200-(75*F):IF F>2.2 THEN 6920
6900 SLOPE(L)=(10+ACC(1))-(10+ACL(L))
6910 IF SLOPE(L)<0 THEN FLAG77=0:RETURN
6920 IF KSX<5 THEN KSX=6
6930 IF KSX>635 THEN FLAG77=0:RETURN
6940 IF FSY<5 THEN FSY=6
6950 IF FSY>185 THEN FSY=184
6960 IF F=2.5 THEN CIRCLE (KSX,FSY),2,1:GOTO 6980
6970 LINE -(KSX,FSY):CIRCLE (KSX,FSY),2,1
6980 IF L<23 THEN F=F-.1:L=L+1:GOTO 6880 ELSE FLAG77=0:RETURN
6990 CLS:END
```

4 Distinct element models of rock and soil structure

P. A. CUNDALL

4.1 Introduction

The **distinct element method** (see Cundall & Strack 1979) was originally conceived as a means to model the progressive failure of rock slopes. For example, Figure 4.1 illustrates a sequence of states calculated for a slope that has planes of discontinuity parallel to the slope face (e.g. steeply dipping foliation planes). The subhorizontal faults permit a rotational mode of failure that is not commonly discussed in the literature.

John Bray has always encouraged development of the distinct element method, but has rightly questioned two aspects that seem to be in need of improvement. These items, the law of motion and the influence of damping on the solution, are discussed below, in response to Bray's questions.

In the final sections of the chapter two examples are presented in which the distinct element method is used to investigate the effect of structure, or fabric, on the behaviour of a rock and soil sample, respectively.

4.2 The difficulties with the distinct element method

4.2.1 The law of motion

The distinct element method is normally used to determine if a rock mass will fail under a given set of applied loads (including gravity), or to calculate the displacements that are accumulated if the system finally stabilizes. The method uses Newton's law of motion to obtain velocities and displacements from unbalanced forces (where m is the mass of the block and the summation term is for all forces applied to the block):

$$\ddot{u} = \frac{1}{m} \Sigma F$$

$$\dot{u} = \int \ddot{u} \, dt \qquad (4.1)$$

$$u = \int \dot{u} \, dt$$

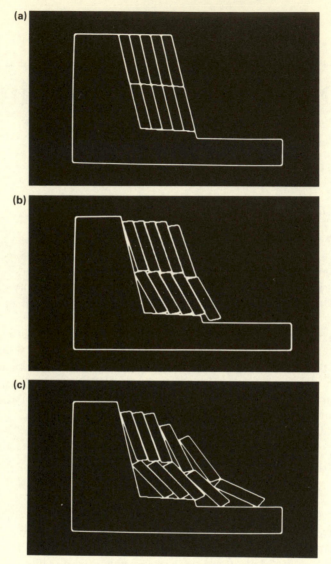

Figure 4.1 Three stages in the failure of a slope, calculated by the distinct element method.

It may be argued that velocities and inertial forces are unimportant in a quasi-static solution method and that a more efficient solution scheme could be devised. At any stage in the solution of a non-linear problem in geomechanics, there exist unbalanced forces; that is, elements (blocks or zones) will not be in perfect equilibrium with their neighbours. The task of the solution scheme is to determine a set of displacements that will bring all elements to equilibrium or, if equilibrium is not possible, to indicate the failure mode. Furthermore, the numerical scheme must satisfy the given

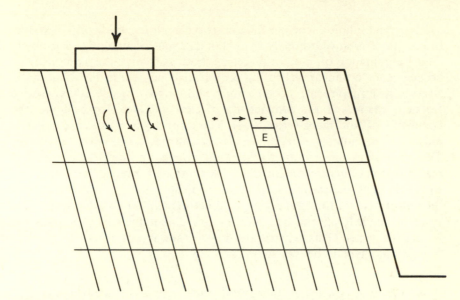

Figure 4.2   Example of the collapse of a footing in rock: element E is carried along by the flow.

constitutive laws for elements and interfaces and follow a path that is
physically possible and reasonably likely to occur in reality.

In order to study the effects of various solution schemes, consider the rock
system shown in Figure 4.2. The example is somewhat contrived, but it
serves to illustrate important characteristics of several solution schemes.
The mode of collapse is not obvious and will probably depend on friction
values. It seems likely that the left-hand region will develop a rotational
mechanism similar to that shown in Figure 4.1. The right-hand region may
behave in a similar fashion, or may simply translate as a single unit. Once the
collapse mechanism has developed and steady-state motion established, a
typical element E experiences no net force; it is carried along with the 'flow'.
A similar thing is noted in a soil mass that is flowing under a footing load (see
Fig. 4.3, in which an analogous soil element E is illustrated).

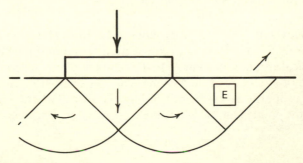

Figure 4.3   Footing collapse in soil: element E is carried along by the flow.

Numerical solution schemes fall into two categories. In the first, a system of equations is solved directly by a standard matrix method, for example Gauss elimination or another direct method. Although this approach can account for non-linearities, using an iterative procedure in which initial stresses or strains are introduced it is not well suited to problems that involve frequent changes in the connectivity of elements (see, for example, the comments of Rodriguez-Ortiz 1974). The difficulty here is that the matrix must be re-formulated every time a connection breaks or a new one is made. The other approach is to use a relaxation scheme, in which the storage and manipulation of matrices are avoided. A common form of relaxation is one where element (block) displacements are adjusted in such a way as to reduce the out-of-balance forces at nodes (or block centroids). In other words, the system is made to approach equilibrium incrementally, while compatibility is always satisfied. This may be done in several ways, of which two (termed 'type A' and 'type B' below) are in common use:

Type A relaxation schemes. In methods such as those of Jacobi and Gauss–Seidel, each node is given a displacement that would bring it to equilibrium if other nodes were to remain fixed. As each node is moved in turn, the equilibrium of previous nodes is destroyed, but after several passes through the system of equations the out-of-balance forces become small, and all nodes are approximately in equilibrium. The method of successive over-relaxation is similar, but the displacements given at each step are larger than those determined for the other two methods. In all methods except Jacobi's the process is path-dependent and therefore to be used with caution when the system is non-linear. Finally, there are variants in which a notional velocity or displacement increment is associated with each node and made proportional to the out-of-balance force. This technique was tried by Cundall (1971: 127) and was also suggested by Bray (1975) as a numerical analogue of the base friction method. Even though velocities are used, the method is identical in form to those previously discussed, since displacement increments are proportional to out-of-balance forces for a constant time step.

Type B relaxation scheme. In dynamic relaxation (Otter *et al.* 1966) the nodes are moved according to Newton's law of motion, which is justified on physical grounds, in contrast to the schemes discussed above, which are justified on the basis of their numerical efficiency. The acceleration of a node is proportional to the out-of-balance force (where the superscripts denote the point in time at which the corresponding variable is evaluated):

$$\ddot{u}^{(t)} = \frac{1}{m} \sum F^{(t)} \qquad (4.2)$$

The equation is integrated twice by central finite differences to produce new velocities and new displacements:

$$\dot{u}^{(t+\Delta t/2)} = \dot{u}^{(t-\Delta t/2)} + \ddot{u}^{(t)} \Delta t \qquad (4.3)$$

$$u^{(t+\Delta t)} = u^{(t)} + \dot{u}^{(t+\Delta t/2)} \Delta t \qquad (4.4)$$

Since an elastic system would continue to oscillate for ever, damping must be provided so that the steady state is approached. In the original form of dynamic relaxation, damping forces are proportional to nodal velocities:

$$F_d^{(t)} = -c\dot{u}^{(t)} \qquad (4.5)$$

The corresponding centred difference equation becomes

$$\dot{u}^{(t+\Delta t/2)} = \frac{\dot{u}^{(t-\Delta t/2)}(1/\Delta t - c/2) + \ddot{u}^{(t)} \Delta t}{(1/\Delta t + c/2)} \qquad (4.6)$$

It should be appreciated that physical time is of no real significance in either of the two approaches summarized above; the objective is to achieve the steady state (whether it is equilibrium or steady flow) with minimum computer effort and in a way that is numerically stable. For example, nodal masses may be scaled for optimum convergence in dynamic relaxation.

Otter *et al.* (1966) point out that dynamic relaxation converges faster in the elastic case than Jacobi iteration when optimum damping is used. Successive over-relaxation on the other hand is shown to converge faster than dynamic relaxation. Although it is undoubtedly true that type A relaxation schemes can be faster than dynamic relaxation for the elastic case, there is a serious difficulty in using them for cases where flow occurs, even if the flow is only transient.

The difficulty is that spurious body forces are introduced by the type A schemes for elements that move with constant velocity (or have a constant displacement increment per iteration step). This can be seen by recalling that elements only move in response to an out-of-balance force:

$$\Delta u \propto \sum F \qquad (4.7)$$

In other words, a force must be exerted on an element in order for its displacement to change. Returning now to the example of Figure 4.2, we see that a typical element E in the right-hand region (the 'flowing' region) acts as if there is a drag force acting on it, i.e. its neighbours must exert a force on it if it is to continue to move at a constant rate. The forces are cumulative for a row of blocks in series, so that the left-hand region can experience a large artificial resistance from the right-hand region, in addition to the real frictional forces. If the rate of movement is large enough, the drag forces may change the mode of failure; in any event the observed collapse load will be rate-dependent, which is an undesirable feature in a numerical model. If the rate of movement is restricted so that drag forces are very small, the solution process becomes very slow, since more iteration steps must be taken for a required displacement to be achieved.

If dynamic relaxation is used to model the same collapse problem, a region that is moving at a steady speed will experience no drag forces if velocity-proportional damping is not used; it is only accelerations that give rise to body forces. In principle, therefore, dynamic relaxation can model collapse problems in a more realistic, efficient manner. In practice there are difficulties associated with time step and damping that must be addressed. As mentioned before, when physical time is unimportant, the element masses can be scaled so the local critical time steps are all equal. For example, small elements can be given inertial masses that are the same as the large elements (of course, the gravitational masses are not affected). In this way the solution process is optimized, so that the convergence time is in the same order of magnitude as type A relaxation schemes. Damping presents a more difficult problem because the usual form of viscous damping (Eqn 4.5) introduces body forces similar in nature to those of the type A relaxation schemes. However, in dynamic relaxation, the body forces can be controlled independently, whereas they are built into the solution process in type A relaxation schemes. Some alternative forms of damping for dynamic relaxation are presented in the next section. Finally, the various relaxation and damping schemes are illustrated in Section 4.2.3 with an example.

4.2.2 Damping

The use of velocity-proportional damping in standard dynamic relaxation involves three main difficulties.

(a) The damping introduces body forces, which are erroneous in 'flowing' regions, and may influence the mode of failure in some cases.
(b) The optimum proportionality constant depends on the eigenvalues of the matrix, which are unknown unless a complete modal analysis is done. In a linear problem, this analysis needs almost as much computer effort as the dynamic relaxation calculation itself. In a non-linear problem, eigenvalues may be undefined.
(c) In its standard form, velocity-proportional damping is applied equally to all nodes, i.e. a single damping constant is chosen for the whole grid. In many cases a variety of behaviour may be observed in different parts of the grid; for example, one region may be failing while another is stable. For these problems, different amounts of damping are appropriate for different regions.

In an effort to overcome one or more of these difficulties, alternative forms of damping may be proposed. In soil and rock, natural damping is mainly hysteretic; if the slope of the unloading curve is higher than that of the loading curve, energy may be lost. The type of damping can be reproduced numerically, but there are at least two difficulties. Firstly, the precise nature of the hysteresis curve is often unknown for complex loading–unloading paths. This is particularly true for soils, which are typically tested with sinusoidal stress histories. Cundall (1976) reports that very different results are obtained when the same energy loss is accounted for by different types of hysteresis loop. Secondly, ratcheting can occur, i.e. each cycle in the oscillation of a body causes irreversible strain to be accumulated. This type of damping has been avoided, since it increases path-dependence and makes the results more difficult to interpret.

Adaptive damping has been described briefly by Cundall (1982). Viscous damping forces are still used, but the viscosity constant is continuously adjusted in such a way that the power absorbed by damping is a constant proportion of the rate of change of kinetic energy in the system. The adjustment to the viscosity constant is made by a numerical servomechanism that seeks to keep the following ratio equal to a given ratio:

$$R = \sum P / \sum \dot{E}_k \qquad (4.8)$$

where P is the damping power for a node, $\dot{E}_k$ is the rate of change of nodal kinetic energy, Σ represents the summation over all nodes. This form of damping overcomes difficulty (b) above, and partially overcomes (a), since as a system approaches steady state (equilibrium or steady flow) the rate of change of kinetic energy approaches zero and consequently the damping power tends to zero. Finally, a new form of damping is proposed here in which the damping force on a node is proportional to the magnitude of the out-of-balance force, with a sign that ensures that vibrational modes are damped, rather than steady motion.

$$F_d \propto |F| \; \text{sgn} \, (\dot{F}) \qquad (4.9)$$

where F is the nodal out-of-balance force. This type of damping is equivalent to a local form of adaptive damping described above. In principle, all of the difficulties reported above are overcome: body forces vanish for steady-state conditions; the magnitude of damping constant is dimensionless and does not depend on properties or boundary conditions; the amount of damping varies from point to point as required. However, it seems that systems are always underdamped when using this scheme, i.e. decaying oscillations are observed.

4.2.3 A numerical example

The various relaxation and damping schemes are compared by modelling a row of blocks resting on a frictional base. Figure 4.4 shows the geometry,

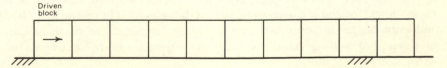

Figure 4.4 Example used in illustrative calculation: ten blocks slide on a rough base.

which is intended to reproduce, in simplified form, the conditions experienced by the right-hand region of Figure 4.2. The blocks can only move in the x direction, and have the following properties:

normal stiffness between blocks	1 unit
shear stiffness between each block and base	1 unit
maximum shear force that can be sustained between each block and the base (i.e. the frictional force)	0.5 units
block mass	1 unit

The left-hand block is pushed either by a force or by moving it at a constant velocity. For a low value of force, the block system comes to equilibrium in a few relaxation steps whichever of the schemes is used. For a force that is large but below that necessary to cause the whole system to slide, the number of relaxation steps becomes much larger, and the different schemes begin to exhibit different characteristics. The differences are accentuated in the extreme case, when all blocks slide at the same speed; this is accomplished by suddenly giving the left-hand block a constant velocity of 1 unit. Figure 4.5 presents a comparison of the following methods and parameters:

(1) Velocity proportional to out-of-balance force; proportionality constant = 1.667.
(2) Dynamic relaxation; standard velocity-proportional damping; damping constant = 0.1.
(3) Dynamic relaxation; adaptive damping – initial damping constant = 0.1; target R-value = 0.8.
(4) Dynamic relaxation; new local damping scheme; proportionality constant = 0.8.

In all the above cases the time-step is 0.2 units. Figure 4.5 shows the reaction force on the left-hand block; it is the force that must be exerted on the block to cause it to move at constant speed. The reaction force is normalized by dividing the calculated force by the maximum total base resistance (5 units in this example).

It is noticed that scheme 1 results in a large overestimate of the driving force; this is in consequence of the erroneous body forces discussed in Section 4.2.1. Scheme 1 also takes a large number of steps to converge, even though the optimum constant for the elastic case is used. Scheme 2 overestimates the force by much less, but an oscillation is superimposed; a larger damping coefficient would reduce the oscillation but increase the error in force. Both schemes 3 and 4 converge to the correct value of force but show oscillations. The test is a severe one in that the velocity is large and is applied suddenly, but the intention is to emphasize the characteristics of the various methods. For smaller velocities, both the erroneous body forces and the oscillations are smaller.

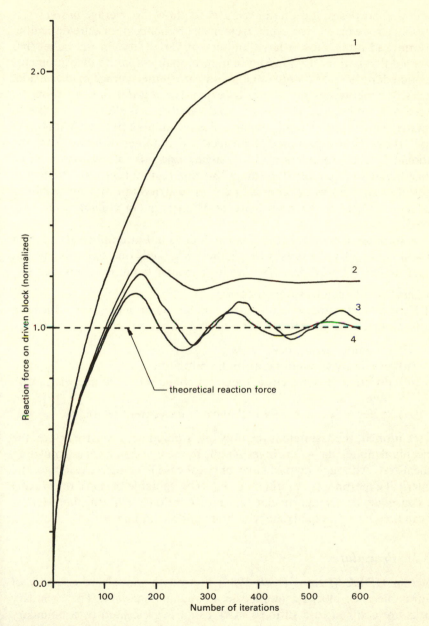

Figure 4.5 Reaction force on left-hand block of Figure 4.4 when it is moved with constant velocity; results are shown for different relaxation schemes.

4.3 Structural effects in a rock mass

For design purposes, it is often necessary to know the average *in situ* stress tensor in a rock mass. However, stress measurements in rock are difficult to perform and often show a large amount of variability. In the belief that observed fluctuations are random in nature, many separate measurements are often done, so that a more reliable estimate of the average may be made. Since stress measurements are always expensive, it is desirable to know the source of the observed fluctuations and whether it is thereby possible to optimize the sampling strategy to minimize the number of tests. With this in mind, Barry Brady proposed a project and obtained funding from the National Science Foundation to investigate one possible source of stress variability and to quantify it, perhaps by using regionalized variable theory (geostatistics). The work reported here was carried out by José Lemos, a graduate student at the University of Minnesota, in collaboration with myself.

It is to be understood that the research does not attempt to account for variability in rock mass stress in a complete way: only one aspect is studied, namely item (iv) below. Stress variability can arise from two main sources:

(a) Instrument and measurement errors.
(b) Natural fluctuations of stress from point to point in the rock, which can be further attributed to at least four causes:

  (i)   inhomogeneous rock;
  (ii) proximity to major faults or discontinuities;
  (iii) differential contraction, creep, or changes in rock properties with time;
  (iv) cycles of tectonic activity that cause movement on joints.

As mentioned, the last item is the only one studied here. Furthermore, the conceptual model that we are investigating by means of numerical simulation is idealized. Although no particular physical case is being represented, the numerical experiments are still believed to be valuable because they reveal the consequences of a particular set of assumptions. With this knowledge, we can then interpret field data with more understanding.

4.3.1 Procedure

Program UDEC (Lemos *et al.* 1985) is used to model an assembly of deformable rock blocks contained in a circular boundary. The boundary blocks are coupled to a stiffness matrix that is generated by a boundary element model of a circular hole in an infinite continuum. In other words, the blocks appear to be embedded in an infinite elastic continuum, which has the same average properties as the rock mass. The formulation is similar to that described by Brady in Chapter 5. Figure 4.6 illustrates the scheme.

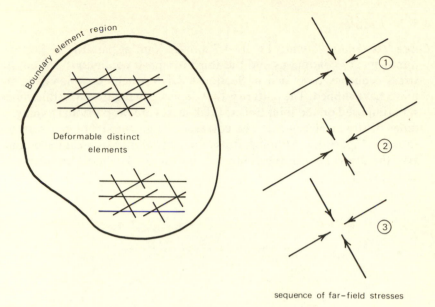

Boundary element region

Deformable distinct elements

sequence of far-field stresses

Figure 4.6   Illustration of boundary element/distinct element interface, and loading sequence of far-field stresses.

Jointing patterns are generated by UDEC by means of a generator that can impart a random deviation to any or all of the generation parameters: spacing, angle, trace length, and gap length. Tests have been performed on many patterns, but only two are reported here. The numerical tests are done by first establishing equilibrium under isotropic stresses; at this stage the joint shear stresses are zero. Then the far-field stress ratio is increased from 1 to 4, with the axis of major principal stress at 30° to the x axis (see Fig. 4.6). Finally, the far-field stress ratio is decreased to unity again. During the second stage, sliding occurs on some joints in the interior part of the sample; joints near the boundary are prevented from sliding by the constraining action of the boundary element region, which is necessarily elastic. Although the boundary causes some non-uniformity, the effect is confined to a strip around the boundary. With either a constant stress or a fixed boundary the effect is much worse. Furthermore, the circular boundary was chosen because it also reduces boundary effects compared to a rectangular boundary. Material properties are summarized as follows:

Young's modulus	50 GPa
Poisson's ratio	0.25
joint friction angle	30°
joint normal stiffness	100 GPa/m
joint shear stiffness	100 GPa/m
radius of sample	25 m

4.3.2   Results

Case 1: continuous jointing. Figure 4.7 shows a jointing pattern in which two joint sets are continuous and the third is almost continuous. When the stress sequence described in Section 4.3.1 has been completed, many joints have slipped. The pattern is that depicted in Figure 4.8 by thick lines superimposed on the joint traces; thicknesses are proportional to magnitudes of slip displacement. The corresponding contours of *xx* stress are shown in Figure 4.9. Although stress concentrations exist near the boundary, the stress field in the centre of the sample is almost uniform and isotropic.

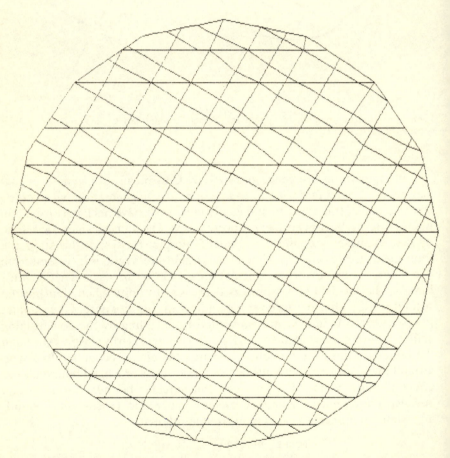

Figure 4.7   'Continuous' jointing pattern used for first numerical experiment on jointed rock.

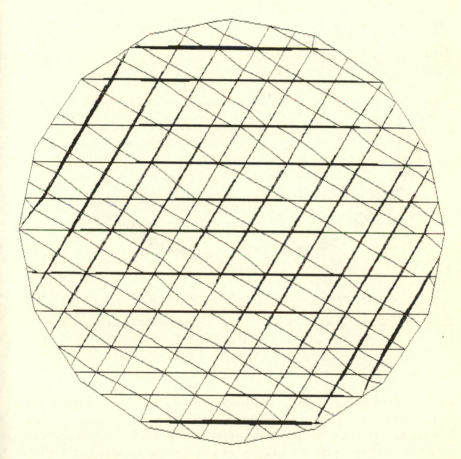

maximum shear displacement = 0.002

Figure 4.8 Thick lines indicate the magnitude of slip that takes place when the sample is loaded and then unloaded deviatorically.

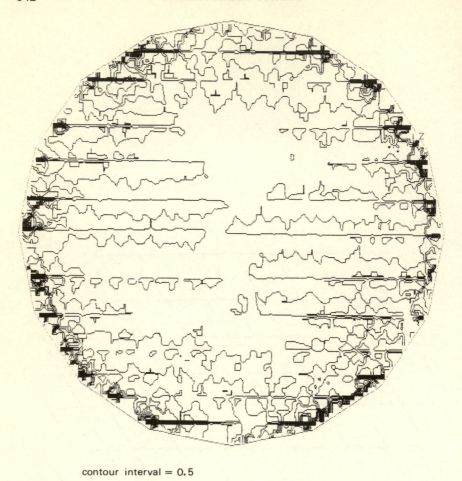

contour interval = 0.5

Figure 4.9 Contours of horizontal stress. Stress concentrations are seen only near the boundary.

Case 2: discontinuous jointing. The numerical experiment described above is repeated, but with the discontinuous joint set shown in Figure 4.10. The resulting slip displacements are illustrated in Figure 4.11, and the distribution of *xx* stress in Figure 4.12. In contrast to case 1, rather large stress concentrations are introduced by the simulated tectonic activity. It is observed that stress concentrations are induced wherever slipping joints terminate at another cross joint. At this point there is a stress jump across the slipping joint: if slip is towards the cross joint, the compressive stress is increased; if slip is away from the cross joint, the compressive stress is reduced.

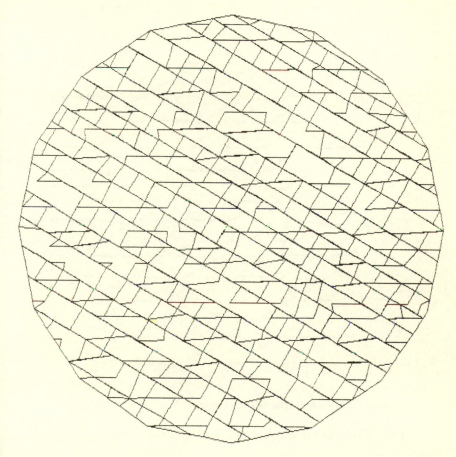

Figure 4.10 Discontinuous jointing pattern used for second numerical experiment on jointed rock.

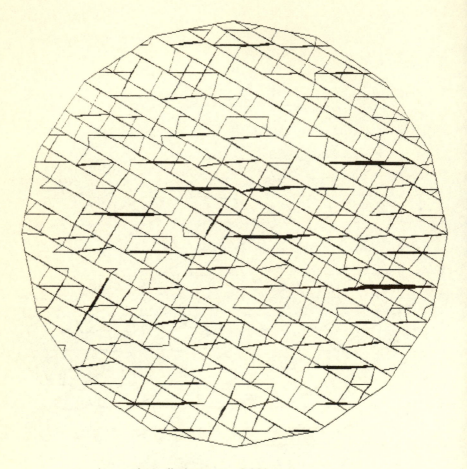

maximum shear displacement = 0.001

Figure 4.11 Slip that occurs during the loading/unloading sequence.

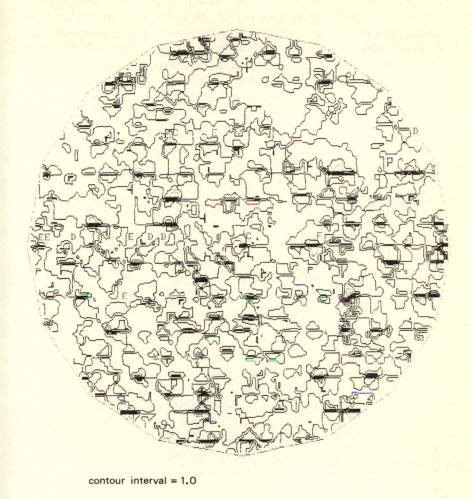

contour interval = 1.0

Figure 4.12 Contours of horizontal stress. The 'locked-in' stresses at the end of the loading sequence are evident.

Case 3: isolated joints. In order to understand the nature of the stress concentrations that are induced around finite length joints, a single fracture is modelled under the same stress path as in the previous cases. Figure 4.13 shows the numerical grid; all joints were prevented from sliding, except the one marked X–X. Again, the stress jump is seen (Fig. 4.14) across the joint tips, even though no cross joint is present. In fact an exact solution exists for the stress concentrations that develop around a single shear crack. For example, Lemos (1986) derives expressions for *xx* stress on the centre line of a crack of half-length *a*, orientated along the *x* axis:

$$\sigma_{xx}^+ = 2\tau_{max}\, x/(a^2 - x^2)^{1/2} \tag{4.10}$$

$$\sigma_{xx}^- = -2\tau_{max}\, x/(a^2 - x^2)^{1/2} \tag{4.11}$$

EE6B

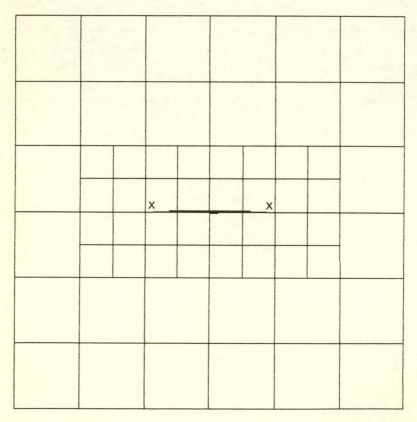

X–X denotes location of joint

Figure 4.13   Numerical grid for a single joint in rock.

EE6B

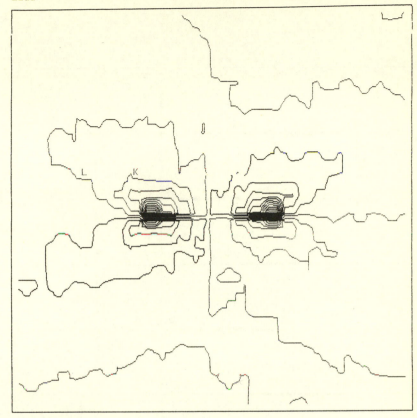

contour interval = 0.5

Figure 4.14   Stress concentrations that arise from the single joint.

where x is the distance from the centre of the crack. Equation 4.10 is for the upper surface for the crack and 4.11 for the lower surface; the sum of the two expressions is zero, which shows that the xx stress vanishes for points on the centre line but away from the crack. Figure 4.14 indicates that the area of high stress concentration is located close to the joint; consequently, it may be possible to estimate the effects of many joints by superposition. To illustrate this possibility, the previous simulation is repeated with multiple parallel joints (see Fig. 4.15). Figure 4.16 shows the computed stress concentrations. It appears that each joint produces approximately the same effect as if it existed alone.

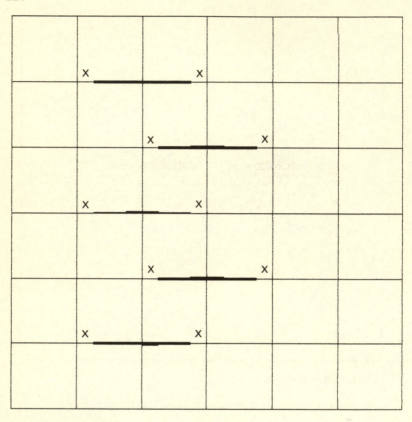

X – X denotes location of joints

Figure 4.15   Multiple, parallel joints.

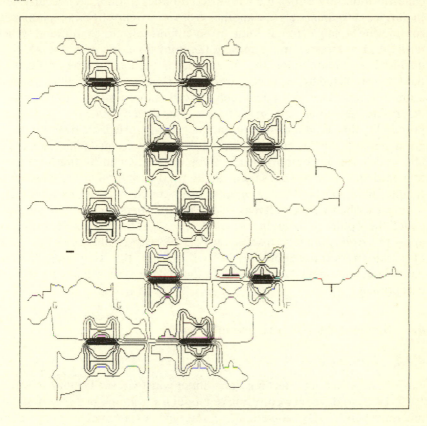

contour interval = 0.5

Figure 4.16 Stress concentrations that arise from the multiple, parallel joints.

4.3.3   Summary and conclusions

The simulations show that a cycle of tectonic activity can cause permanent stress concentrations to be induced in a jointed rock mass. Such stress concentrations are associated with discontinuous jointing and occur where joint traces terminate, e.g. on another cross joint. The effect is explained by considering a single rough joint in an infinite elastic continuum. When far-field shear stresses are increased, the joint slides but does not return to its initial state upon unloading. The resulting permanent stress concentrations are predicted by an exact solution for a shear crack. If eligible cracks in a jointed rock mass are quite widely spaced, superposition can be used to determine the overall pattern of stresses. It is then possible to calculate the probability that a random measurement of stress will deviate by more than a given amount from the mean stress: the probability is equal to the sum of the areas enclosed by contours of the given stress divided by the total area. Standard statistical methods then can be used to determine the number of samples that must be taken in order to estimate the average stress to within some prescribed accuracy. It may also be possible to make recommendations about the optimum location of sample points. The preceding comments apply to a strictly limited subset of the possible ways in which measurements of *in situ* stress can exhibit variability; however, the study indicates that natural fluctuations in stress are to be expected in circumstances that are not exceptional.

4.4   Structural effects in a granular soil

4.4.1   Introduction

For several years, a project has been under way (e.g. see Cundall & Strack 1979, 1983) to develop a constitutive model for granular material based on micromechanics. By conducting numerical experiments on granular assemblies (see Fig. 4.17) it is possible to study in detail the mechanisms at the particle level that are responsible for stress–strain relationships observed in the laboratory or in the field. A continuum constitutive model will then be built on the basis of these observed mechanisms; it is hoped that such a model will be more satisfactory than models derived from intuitive assumptions and curve-fitting, which are common approaches.

A complete constitutive model has not yet been built, although certain aspects of behaviour are already reproduced by average-particle models (Strack & Cundall 1984). One characteristic of granular material that is important, and is now relatively well understood at the particle level, is the influence of structure, or fabric, on the memory property of granular materials. Some results have already been reported for the case of two-dimensional disk assemblies (Cundall & Strack 1983). For example, Figure 4.17 illustrates a numerical shear box test done on an assembly of disks; the

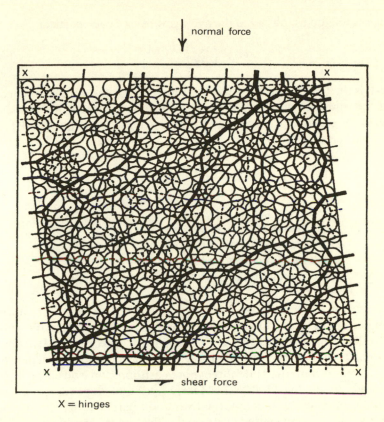

Figure 4.17 Example of numerical experiments on 2D disks. Magnitude and direction of contact forces are given by thickness and angle of lines.

thick lines indicate the magnitude and direction of contact forces. The test was done in an attempt to reproduce results from a similar physical test.

In Section 4.4.4 the results from three-dimensional experiments will be presented, but before that it is necessary to review the development of stress partitions and to explain how numerical tests on three-dimensional assemblies are carried out. It should be pointed out that all tests and theory referred to here are restricted to circular or spherical particles, although the distinct element method is equally applicable to angular particles. When a satisfactory theory for irregular distributions of circular particles has been formulated, we shall feel justified in moving to the case of particles of complex shape.

4.4.2 Stress partitions

In this section the Einstein summation convention applies to repeated indices which take values from 1 to 3. The average stress in an assembly of particles is related as follows to the average stress of each particle:

$$\bar{\sigma}_{ij} = \frac{1}{V} \sum_p \sigma_{ij}^{(p)} V^{(p)} \tag{4.12}$$

where V is the total volume (including voids), $V^{(p)}$ is the volume of particle p, $\sum_p$ is the summation over all particles. The average stress of a particle is related to the contact forces acting on the particle, so that Equation 4.12 for circular particles becomes

$$\bar{\sigma}_{ij} = \frac{1}{V} \sum_p R^{(p)} \sum_c n_i F_j \tag{4.13}$$

where

$R^{(p)}$	is the radius of particle p
n_i	is the unit normal vector at contact c
F_i	is the force vector at contact c
$\sum_c$	is the summation over all contacts on particle p

The contact force can be decomposed into a shear component and a normal component, so that the average stress tensor can be expressed as the sum of two partitions: one that depends on the shear force and one that depends on the normal force. Following Cundall and Strack (1983), the normal partition can be further decomposed into three more partitions: an isotropic part that corresponds to the mean stress, a fabric part that depends on angular distribution of contacts, and a normal variation part that depends on how normal forces vary in magnitude with angle. The shear partition is

$$\sigma_{ij}^{(s)} = \frac{1}{V} \sum_p R^{(p)} \sum_c (n_i F_j - F_k n_k n_i n_j) \tag{4.14}$$

and the fabric partition is

$$\sigma_{ij}^{(f)} = \frac{1}{V} \sum_p V^{(p)} \sigma_0^{(p)} \left(\frac{\delta_{kk}}{m} \sum_c n_i n_j - \delta_{ij} \right) \tag{4.15}$$

where $\sigma_0^{(p)}$ is the mean stress of particle p, m is the number of contacts on particle p. During a numerical test on an assembly of spheres, the average stress is computed according to Equation 4.12 and the shear and fabric partitions according to Equations 4.14 and 4.15. The partitions may be regarded as internal variables that cannot be determined by conventional measurements at the boundary of a sample. Figure 4.18 illustrates how each stress partition corresponds to a particular way in which deviatoric stress can arise from changes in an initially isotropic force distribution on an average particle. It is shown in Section 4.4.4 that the overall deviatoric stress may be zero, but two internal stress partitions may be equal and opposite.

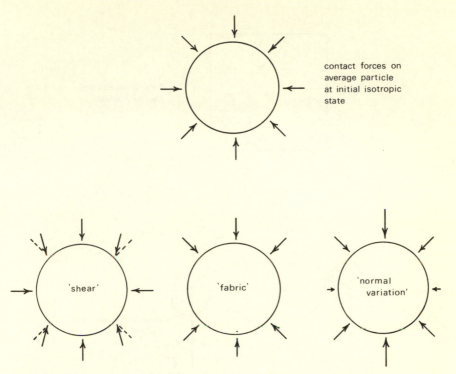

contact forces on
average particle
at initial isotropic
state

'shear'

'fabric'

'normal
variation'

Figure 4.18 The sketch illustrates three different ways in which the contact force distribution on an average particle can change in order to give rise to a deviatoric stress; each way corresponds to a stress partition.

4.4.3   *Numerical tests in three dimensions*

It was discovered from tests in two dimensions that boundaries influence the behaviour of a granular assembly; a force boundary offers more freedom to particles than they experience normally, and a velocity boundary provides more constraint. A disturbed region is seen to penetrate to a depth of a few particle diameters from a boundary. In three dimensions the boundary effect will be more serious for the same number of particles in a sample because there are fewer particles across any dimension in the sample. It was decided therefore to run all tests without boundaries, by using periodic assemblies. In this way the boundary effect is exchanged for effects of periodicity. It is believed that the latter are less disruptive. In the two-dimensional periodic space illustrated in Figure 4.19, particles that move outside the left limit of the space reappear at a corresponding point at the right limit. The upper and lower spatial limits are similarly connected topologically. A good visualiza-tion of periodic spaces is given by Thurston and Weeks (1984). Since the sample has no boundaries, strain is applied to the particle assembly by

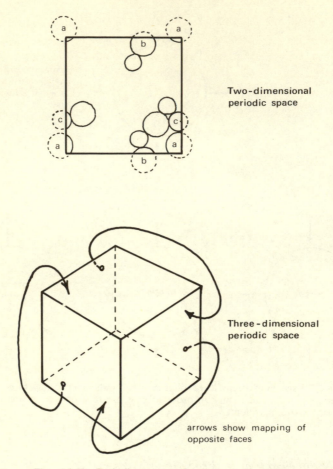

Two-dimensional
periodic space

Three-dimensional
periodic space

arrows show mapping of
opposite faces

Figure 4.19   Periodic two- and three-dimensional spaces.

distorting the periodic space itself, and by changing its volume. In order to visualize how this process works, consider the example of a number of two-dimensional disks attached without friction to the surface of a spherical balloon. As the balloon contracts, the disks become closer together, that is, although disks have no absolute velocity, each disk perceives its neighbours to move closer with a velocity that is proportional to their distance apart. When disks touch, and forces arise between them, they acquire a real velocity relative to the two-dimensional space that they inhabit. This new velocity is superimposed on the relative velocity between disks that comes from the strain rate of space. In three dimensions, a similar analogy can be made, but here the particles must be imagined to exist on the three-dimensional 'surface' of a four-dimensional body.

In program TRUBAL, which models assemblies of spheres, the effects of a periodic space that changes in shape and size are incorporated as follows.

(a) Particles that leave one side of the periodic cell reappear at the corresponding position on the opposite face (see the lower sketch in Fig. 4.19). A particle at the edge of the cell can interact with particles on the opposite face of the cell. In order to detect if two particles are in contact across the cell, the periodic dimension is subtracted from the difference in their coordinates.

(b) At each time step, a displacement increment due to the strain rate of periodic space is added to the displacement increment calculated from the real particle velocity. The strain rate of space is taken to be independent of position, and is denoted by $\dot{e}_{ij}^{(s)}$.

$$\Delta u_i := \Delta u_i + \dot{e}_{ij}^{(s)} x_j \Delta t \qquad (4.16)$$

(c) The relative approach velocity of two particles is found by summing their relative real velocities and a velocity related to the strain rate of the periodic space:

$$\dot{u}^{(\text{rel})} = \dot{u}^{(\text{real})} + \dot{e}_{ij}^{(s)} z_j^{(\text{diff})} \qquad (4.17)$$

where $z_j^{(\text{diff})}$ is the vector distance between centres.

Following the scheme described, the assembly of spheres is seen to have certain characteristics. Particles are moved by the periodic space, without force or velocity, to the positions they would have in a uniform strain field; they only acquire real velocities when contact between neighbours disturbs them. These velocities, then, only arise when deviations from a uniform strain field are dictated by the discrete nature of the assembly. Out-of-balance forces and damping forces therefore tend to be much smaller than those produced in a test done in the conventional way. Consequently, a test can be done faster on a periodic assembly, for the same level of accuracy. Since there is no location in the assembly that is distinct from any other location (in contrast to a conventional test, in which boundary particles are special), the compacted assembly is statistically uniform.

4.4.4   Test procedure and results

A random number generator is used to place spheres within a cube of side 200 units; no particle is initially allowed to touch any other. For the numerical experiment described here, 150 particles are used: 100 of radius 10 units and 50 of radius 15 units. The initial, random state is termed 'state A', and is illustrated by the computer-drawn picture of Figure 4.20. The periodic space is then contracted isotropically until its cubic dimension is 119.1 units. During this contraction, the friction coefficient is 1.0, but after compaction it is set to 0.05 for 600 time steps in order to achieve an average stress tensor that is more isotropic. Figure 4.21 shows the compacted state, termed 'state B'. The lines drawn on particle surfaces are reference lines that will make visible any rotation that takes place during subsequent loading and

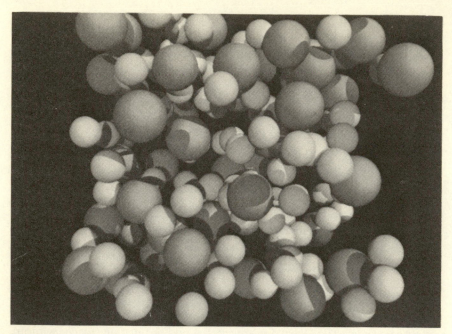

Figure 4.20 Computer-generated picture of the initial distribution of sphere (state A). There are two sizes of particle, and no particle touches another.

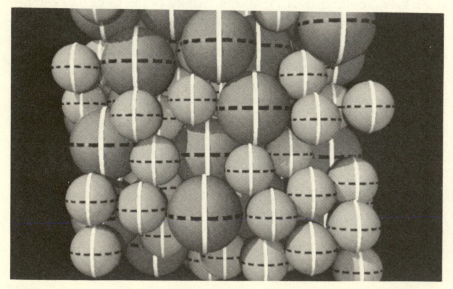

Figure 4.21 Compacted state of spheres before deviatoric loading has been applied. The black and white bands on this computer-drawn picture are markers to indicate rotations. State B is shown.

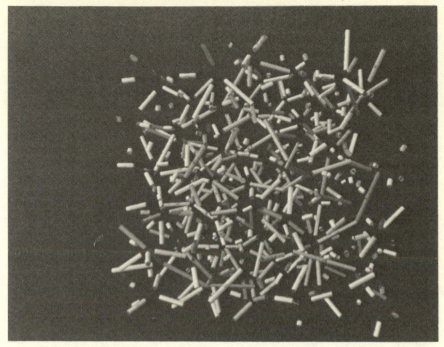

Figure 4.22 'Rods' are drawn by computer to represent in magnitude and direction the normal contact forces. The picture corresponds to state B (of Fig. 4.21).

unloading. Figure 4.22 illustrates the contact forces present in state B: a rod is drawn at each contact point, with a length that is proportional to the normal contact force; shear forces are not visible because of their small magnitude. It may be noted that the forces are orientated more or less at random.

A shear test is then done on the assembly by reducing the x dimension of periodic space and increasing the y dimension while keeping the mean isotropic stress constant at 1.5×10^5 units. Since it is not possible to control the stress directly, a numerical servomechanism is used to adjust the volumetric strain rate in response to errors in the measured isotropic stress (see Cundall & Strack 1983 for details on measurement of the stress tensor in a granular assembly). Material properties used for the test are

friction coefficient	1.0
normal stiffness (linear)	2.0×10^8
shear stiffness (linear)	1.0×10^8

Other tests, not reported here, have been done with non-linear contact stiffnesses, according to the Hertz theory. The measured stress–strain curve for the numerical test is presented in Figure 4.23, which also gives the curve of volume change. At the point denoted as state C, the sample was unloaded

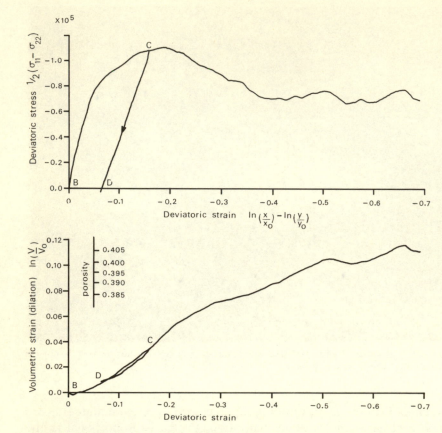

Figure 4.23   Stress–strain and dilatation curves for the numerical test on spheres. A single unloading path to zero deviatoric stress is shown.

to zero deviatoric stress (the subsequent continuation of loading is not affected by this, since state C is saved on file for later re-start). Figure 4.24 illustrates the configuration of particles at state C, and Figure 4.25 shows the corresponding contact forces. The shear forces are now visible as short bars that cross the longer, normal-force bars. Figure 4.26 shows an alternative plot of contact forces in which bars now denote total contact forces and discs are drawn tangentially to each contact point. It is noted that the largest forces in state C are aligned approximately in the direction of major principal strain.

The total deviatoric stress plotted in Figure 4.23 may be decomposed into several partitions, as explained in Section 4.4.2. Figure 4.27 shows the loading–unloading path B–C–D plotted in terms of the stress partitions. The shear partition (derived from shear contact forces) is seen to build up rapidly to a limit, and decrease rapidly on unloading (since sliding ceases

Figure 4.24 State C of the sphere assembly after deviatoric loading has been applied. Note that significant rotations have occurred, and that some particles have migrated across the periodic space.

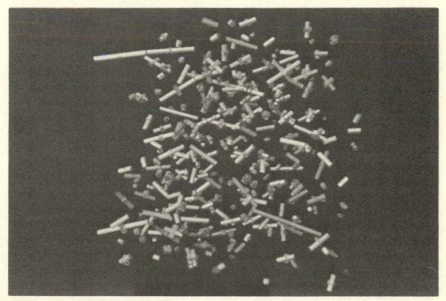

Figure 4.25 Contact force distribution of state C. The normal forces are larger in the x direction (major principal strain), and shear forces are now visible as small bars crossing the longer, normal force bars.

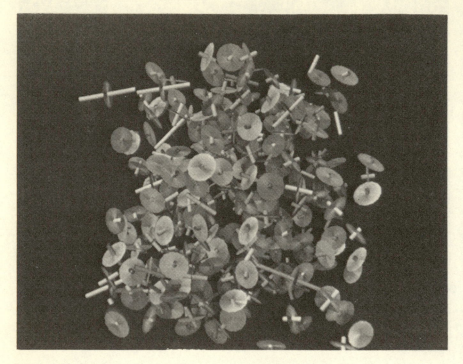

Figure 4.26 An alternative type of computer plot of contact forces. The rods now represent total contact forces, and the disks are drawn to be tangential to contact points.

immediately upon unloading). In contrast, the fabric partition (which embodies changes in contact distribution) continues to build up beyond the point at which the shear partition saturates. Upon unloading, the fabric changes at a slower rate than the shear forces. Consequently, when the total deviatoric stress returns to zero, a negative shear partition is almost balanced by a positive fabric partition; the residue is accounted for by the normal variation partition. This phenomenon constitutes a memory effect: the measured deviatoric stress at D is zero, but internally the state is different from the starting state of B.

The same change in fabric is also responsible for another important effect: increasing dilatation during shearing. During deviatoric loading, the number of contacts in a sample decreases, sometimes by as much as 20%. Contacts that are lost have normals that are generally in the direction of minor principal strain, although a few contacts are added in the major principal direction. The fabric change reflects this changing distribution of contacts. If a granular assembly is viewed instantaneously as a matrix of pin-jointed bars, it is evident that the incremental volume change will only be zero if the areas (or domains) enclosed by bars are, on average, not orientated in any

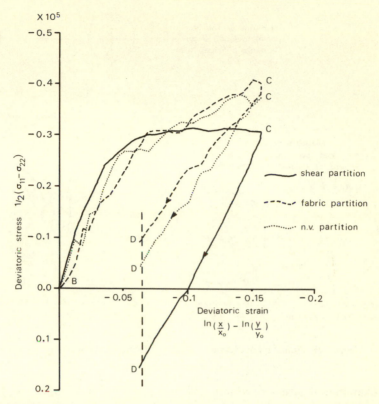

Figure 4.27   Stress partitions are plotted for the path B–C–D (of Fig. 4.23). The sum of the partitions is the measured deviatoric stress. Note that at state D (zero deviatoric stress) the shear and fabric partitions have opposite sign.

particular direction. When contacts (bars) are deleted in the minor principal strain direction, the domains become, on average, elongated in the major principal strain directions. By considering a single example of such a domain, it can be seen that its volume will increase when a compressive strain increment is applied in its long axis. This effect is illustrated in summary in Figure 4.28. The rate of dilatation is expected to be related to the anisotropy of fabric, which is brought about by deletion of contacts in a preferential direction. Figure 4.28 also illustrates the memory effect described earlier, which also is related to changes in fabric.

4.4.5   Conclusions

Some internal mechanisms have been observed when conducting numerical experiments on granular assemblies; and the consequences for macroscopic behaviour have been demonstrated. It is anticipated that the effects can be incorporated in continuum constitutive models for granular material; for

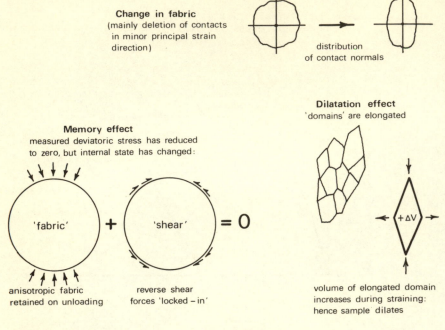

Figure 4.28   Summary of the consequences of fabric change in a granular assembly.

example, if stress partitions are taken as internal variables, it may be possible to reproduce the memory effect more realistically than by using yield surfaces.

4.5   Summary

Some difficulties with the distinct element method have been examined and explanations and possible remedies suggested. The effects of fabric (or structure) in rock and soil have been studied with the distinct element method. Apart from demonstrating that numerical experiments can be useful aids to understanding, the results show that the microstructure of soil and rock can have a profound influence on the overall behaviour of the material.

References

Bray, J. W. 1975. *Methods of analysing discontinua, in situations where slip and separation may produce significant displacements*. Rock Mechanics Technical Note No. 1, Interdepartmental Rock Mechanics Project, Imperial College, London, July.

Cundall, P. A. 1971. *Measurement and analysis of accelerations in rock slopes*. PhD thesis, Imperial College, London.

Cundall, P. A. 1976. Explicit, finite difference methods in geomechanics. In *Numerical methods in geomechanics*, C. S. Desai (ed.), **1**, 132–50. New York: American Society of Civil Engineers.

Cundall, P. A. 1982. Adaptive density-scaling for time-explicit calculations. *Proc. 4th int. conf. numerical methods in Geomechanics*, Edmonton, Z. Eisenstein (ed.), **2**, 23–6. Rotterdam: Balkema.

Cundall, P. A. and O. D. L. Strack 1979. A discrete numerical model for granular assemblies. *Géotechnique* **29**, 47–65.

Cundall, P. A. and O. D. L. Strack 1983. Modeling of microscopic mechanisms in granular material. In *Materials: new models and constitutive relations*, Proc. US–Japan seminar on new models and constitutive relations in the mechanics of granular material, J. T. Jenkins and M. Satake (eds). Amsterdam: Elsevier.

Lemos, J. V. 1986. PhD thesis, in preparation.

Lemos, J. V., R. D. Hart and P. A. Cundall 1985. A generalised distinct element program for modelling jointed rock mass. In *Fundamentals of rock joints*, Stephansson (ed.), 335–43. Luleå: Centak Publishers.

Otter, J. R. H., A. C. Cassell and R. E. Hobbs 1966. Dynamic relaxation. *Proc. Instn Civ. Engrs* **35**, 633–56.

Rodriguez-Ortiz, J. M. 1974. *Estudio del comportamiento de medios granulares heterogeneos mediante modelos discontinuos analogicos y matematicos*. PhD thesis, Universidad Politecnica de Madrid.

Strack, O. D. L. and P. A. Cundall 1984. *Fundamental studies of fabric in granular materials*. Interim report concerning NSF Grant CEE-8310729, Department of Civil and Mineral Engineering, University of Minnesota.

Thurston, W. P. and J. R. Weeks 1984. The mathematics of three-dimensional manifolds. *Scient. Am.* **251**(1), 108–20.

5 Boundary element and linked methods for underground excavation design

B. H. G. BRADY

5.1 Introduction

Underground excavations are designed to perform a prescribed duty or function, such as transport of people and materials, transfer of air or water, or provision of safe and secure work places for mine staff. In each case, the design objective is to ensure that displacements of rock around the excavation are compatible with the performance of specified engineering activities within it.

On the basis of the displacements induced by excavation, it is possible to distinguish four conceptual models of a rock mass, as illustrated in Figure 5.1. The displacement field may be continuous throughout the near field of the excavation, as illustrated in Figure 5.1a. For the rock mass structure shown in Figure 5.1b, within large discrete regions of the near-field rock the displacement field is continuous. These regions are separated by planes of weakness on which slip or separation can occur. When, on the scale of the excavation, a rock mass is frequently jointed, as shown in Figure 5.1c, slip and separation on joints and rigid-body translation and rotation of blocks may determine near-field displacements. For the case of a frequently and randomly fractured rock mass, illustrated in Figure 5.1d, the displacement field is pseudo-continuous. An ubiquitous joint, elastoplastic model may be appropriate for this rock mass condition.

It is clear that, because of the significantly different modes of response for the rock mass conditions illustrated in Figure 5.1, different analytical or computational schemes are required for design analysis for the various cases. Some preceding chapters have been concerned with various analytical and computational procedures for the analysis of stress and displacement

164

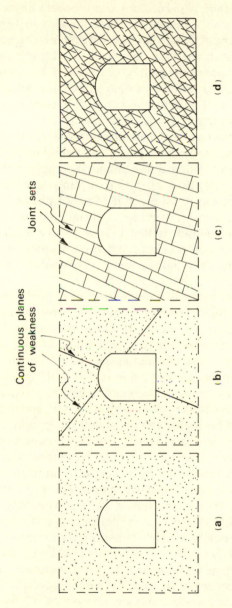

Figure 5.1 Conceptual models relating rock structure and rock response to excavation.

around underground excavations. The following discussion is intended to define appropriate schemes for each of the conceptual models illustrated in Figure 5.1.

The **boundary element method** is particularly useful where linear elastic behaviour can be assumed for a rock mass, or elastic domains are separated by continuous planes of weakness. Thus, the classes of problems represented by Figures 5.1a and b are analysed most conveniently with this method. Several versions and applications of the method are described in subsequent sections. They are characterized by ease of data submission for analysis, relatively small central memory requirement for analysis of complex excavation geometries, and general computational efficiency.

In the discussion of the **distinct element method** of analysis (Cundall 1986), it was clear that the technique provides a powerful method of analysis of problems for rock masses expressing discontinuous or highly non-linear behaviour. This may involve slip and separation at joints and rigid-body translation and rotation of blocks in a jointed medium or plastic flow in a continuous medium. The intrinsic disadvantage of the method, as is also the case with the **finite element method**, is that it is necessary to discretize completely the interior of the problem domain and to specify for that domain both an arbitrary external boundary and the boundary conditions that apply there.

There are clear advantages in developing linked or coupled algorithms which model, with either distinct elements or finite elements, a non-linear domain embedded in an infinite or semi-infinite elastic body modelled with boundary elements. These include, firstly, elimination of uncertainties associated with the conditions to be applied at the outer boundary of the discretized d.e. or f.e. domain. Usually, this region will represent the near-field rock around an excavation. Secondly, when an excavation is developed in a rock mass, far-field rock is subject to only small induced displacements and stresses. This region can be modelled appropriately and conveniently using boundary elements. Thirdly, zones of complex constitutive behaviour in a rock stucture are usually small and localized, so that only these zones require the analytical versatility conferred by finite elements or distinct elements. The implied reduction in size of the zones with a modelled capacity for complex constitutive behaviour again favours computational efficiency. Finally, it is possible to represent the performance of support and reinforcement in near-field rock by special elements interacting with either distinct elements or finite elements. Taking account of all these factors, it is clear that linkage of algorithms representing the key components of rock mass response can produce conceptually valid and computationally efficient schemes for the design of excavations, and any support or reinforcement required within them. In terms of the models of rock mass response illustrated in Figure 5.1, a linked d.e.–b.e. scheme is appropriate for the model in Figure 5.1c and a linked f.e.–b.e. scheme for Figure 5.1d.

In the following discussion, the development of various boundary element and linked algorithms is described, and some applications of them are presented. Unless otherwise stated or implied, the usual geomechanics convention is used for stresses and displacements, with compressive normal stresses positive and displacements in the positive direction of the coordinate axis reckoned as positive. Summation is implied on repeated subscripts and superscripts. The range of summation is from 1 to 2 unless otherwise indicated.

5.2.   Boundary element method for an elastic continuum

5.2.1   *Principles of the boundary element method*

The essence of the boundary element method is definition and solution of a problem entirely in terms of surface values of traction and displacement. For **semi-infinite** and **infinite body** problems, the principle of superposition is employed explicitly in developing the solution procedure, making it most applicable to elastic media. The following discussion is concerned with so-called direct formulations of the boundary element method. It follows the description provided by Cruse (1969), Watson (1979), Brady (1979), and Crotty and Wardle (1985), among many others.

Figure 5.2a shows the cross section of an excavation in an infinite, elastic medium, subject to stresses p_{ij} at infinity. The surface S is traction free after excavation and subject to displacements u_1, u_2 induced by excavation. This can be treated as the superposition of the two problems shown in Figures 5.2b and c: a continuous, uniformly stressed body, and the surface S subject to excavation-induced surface tractions t_α and displacements u_α ($\alpha = 1, 2$). Since the final surface is traction free, induced tractions at any point Q on the surface are given by

$$t_i(Q) = -p_{ij}n_j(Q) \tag{5.1}$$

where $n_j(Q)$ is the unit outward normal to the surface.

Betti's reciprocal work theorem is used to construct a relation between surface tractions and displacements at point Q by considering the fictitious case shown in Figure 5.2d of a unit line load applied at another point P, in directions i ($i = 1, 2$), on the surface S:

$$c_{ij}(P)u_j(P) + \int_S T_{ij}(P, Q)u_j(Q) \, dS = \int_S U_{ij}(P, Q)t_j(Q) \, dS \tag{5.2}$$

In this expression, for doubly subscripted variables, the first subscript represents the direction of the applied fictitious load, and the term in c_{ij} takes account of the singularity at P. In seeking a solution for unknown surface displacements, the procedure is to rewrite the boundary constraint equation

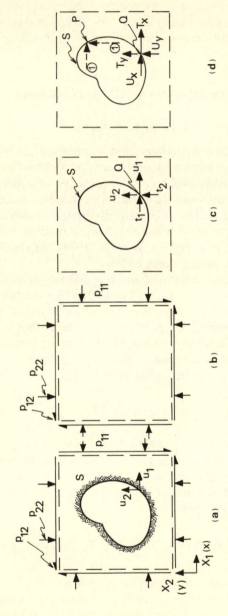

Figure 5.2 (a), (b) and (c) Problem definition for boundary element analysis; (c) and (d) surface loads for development of boundary integral equation.

(i.e. Eqn 5.2) in terms of surface integrals over the set of elements into which the surface is divided, i.e.

$$c_{ij}(P)u_j(P) + \sum_{k=1}^{n} \int_{S_k} T_{ij}(P, Q)u_j(Q) \, dS_k = \sum_{k=1}^{n} \int_{S_k} U_{ij}(P, Q)t_j(Q) \, dS_k \quad (5.3)$$

where n is the number of boundary elements. Equation 5.3 is solved by suitable description of the geometry of each element S_k and by assuming some particular variation of traction and displacement with respect to the element geometry. For numerical integration using Gaussian quadrature, element geometry is represented in terms of a local coordinate ξ. Equation 5.3 becomes

$$c_{ij}(P)u_j(P) + \sum_{k=1}^{n} \int_{S_k} T_{ij}(P, \xi)u_j(\xi)J_k(\xi) \, d\xi = \sum_{k=1}^{n} \int U_{ij}(P, \xi)t_j(\xi)J_k((\xi) \, d\xi \quad (5.4)$$

where

$$J_k(\xi) = \left[\left(\frac{dx}{d\xi}\right)^2 + \left(\frac{dy}{d\xi}\right)^2 \right]^{1/2}$$

is the Jacobian for the coordinate transformation.

5.2.2   Linear isoparametric elements

The surface S may be approximated by a set of linear elements, as shown in Figure 5.3a. A representative element, with nodes 1 and 2 and intrinsic coordinate ξ, is shown in Figure 5.3b. Linear interpolation of position coordinates for points between the nodes is defined by the equations

$$\begin{aligned} x(\xi) &= x^1 N^1(\xi) + x^2 N^2(\xi) = x^p N^p(\xi) \quad -1 \leqslant \xi \leqslant 1 \\ y(\xi) &= y^1 N^1(\xi) + y^2 N^2(\xi) = y^p N^p(\xi) \end{aligned} \quad (5.5)$$

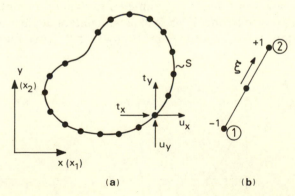

(a)   (b)

Figure 5.3   Discretization of a problem surface into elements defined by surface nodes (a) and a linear boundary element (b).

Here x^p, y^p are nodal coordinates, $x(\xi)$, $y(\xi)$ are the coordinates of a point within S_k, and $N^p(\xi)$ are the linear shape functions given by

$$N^1(\xi) = \tfrac{1}{2}(1 - \xi)$$
$$N^2(\xi) = \tfrac{1}{2}(1 + \xi)$$

(5.6)

In the **linear isoparametric formulation**, variations of traction and displacement are assumed to be defined by the same interpolation functions as the element geometry, i.e.

$$t_i(\xi) = t_i^1 N^1(\xi) + t_i^2 N^2(\xi) = t_i^p N^p(\xi)$$
$$u_i(\xi) = u_i^1 N^1(\xi) + u_i^2 N^2(\xi) = u_i^p N^p(\xi)$$

(5.7)

5.2.3   Quadratic isoparametric elements

For the element and shape functions shown graphically in Figure 5.4, element geometry is approximated by

$$x(\xi) = x^1 \bar{N}^1(\xi) + x^2 \bar{N}^2(\xi) + x^3 \bar{N}^3(\xi) = x^p \bar{N}^p(\xi) \qquad p = 1, 2, 3, \quad -1 \leqslant \xi \leqslant 1$$
$$y(\xi) = y^1 \bar{N}^1(\xi) + y^2 \bar{N}^2(\xi) + y^3 \bar{N}^3(\xi) = y^p \bar{N}^p(\xi)$$

(5.8)

where the shape functions are given by

$$\bar{N}^1(\xi) = -\tfrac{1}{2}\xi(1 - \xi)$$
$$\bar{N}^2(\xi) = (1 - \xi^2)$$
$$\bar{N}^3(\xi) = \tfrac{1}{2}\xi(1 + \xi)$$

(5.9)

For an isoparametric formulation, displacements are defined in terms of nodal displacements and the element local coordinate by

$$u_i(\xi) = u_i^p \bar{N}^p(\xi) \qquad p = 1, 2, 3, \quad -1 \leqslant \xi \leqslant 1$$

(5.10)

To maintain compatibility between displacement and traction, the order of functional variation of traction over an element should preferably be one less than that for displacement. The value of maintaining this compatibility has been demonstrated by Yeung and Brady (1982). They found that it is

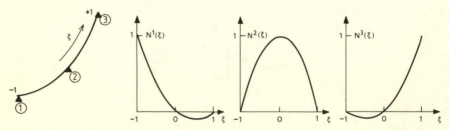

Figure 5.4   Element geometry and shape functions for a quadratic boundary element.

necessary to consider the three-noded element in two sections, with local coordinates α, β for each section. Variation in tractions is then described by

$$t_i(\alpha) = t_i^1 N^1(\alpha) + t_i^2 N^2(\alpha) = t_i^p N^p(\alpha) \qquad -1 \leqslant \alpha < 0$$

$$t_i(\beta) = t_i^2 N^1(\beta) + t_i^3 N^2(\beta) = t_i^{p+1} N^p(\beta) \qquad 0 \leqslant \beta \leqslant 1$$

(5.11)

5.2.4   Solution of boundary constraint equation

Once the form of the element geometry and functional variation is defined, as in Equations 5.5–5.11, and if point P is taken as a boundary node, all the data required to evaluate the integrals in Equation 5.4 are available. The free terms c_{ij} are evaluated by the method due to Cruse (1969), which takes account of the need for force equilibrium over any closed surface in a medium under static load. By considering each surface node in turn, evaluating all the integrals of the kernel-shape function products implied in Equation 5.4, and combining the integrals from adjacent elements for appropriate nodes, a set of $2n$ simultaneous equations is obtained, written as

$$\mathbf{Tu} = \mathbf{Ut} \qquad (5.12)$$

In this equation, $\mathbf{u}$ lists the unknown nodal displacements, $\mathbf{t}$ the known nodal tractions, and the square matrices $\mathbf{T}$ and $\mathbf{U}$ are fully populated and diagonal dominant. Multiplying out the right-hand side of Equation 5.12 produces

$$\mathbf{Tu} = \mathbf{v} \qquad (5.13)$$

which can be solved directly for the nodal displacements using Gaussian elimination or transpose elimination (Wassyng 1982).

After solving for the nodal displacements, the directional derivative of the tangential component of boundary displacement can be determined numerically. This can be used, in conjunction with known boundary tractions, to calculate the tangential component of boundary stress, using the appropriate stress–strain relations.

When the nodal displacements and tractions are known, Equation 5.3, with $u_j(P)$ replacing $c_{ij}(P)u_j(P)$, can be used to calculate displacements at any interior point in the medium. The various directional derivatives of Equation 5.3 are used to establish expressions for strain components in the interior of the medium in terms of nodal parameters. From these, stress components at interior points can be calculated using the appropriate stress–strain relations.

The performance of the quadratic, isoparametric code was assessed using the problem geometry illustrated in Figure 5.5. The discretization of the boundary is seen to be relatively coarse, consisting of only eight elements. The choice of a uniaxial stress field ensured that relatively high gradients existed in boundary stress. Table 5.1 compares, for the b.e. scheme and the Kirsch solution, induced displacements at the nodes and total stresses at the

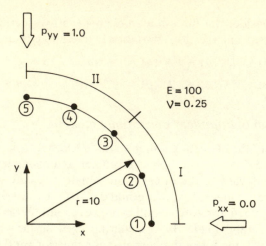

Figure 5.5   Problem geometry for a demonstration of the performance of the b.e. code.

Table 5.1   Comparison of boundary element results with Kirsch's solution.

Induced nodal displacements

Node no.	Computed		Analytical	
	u_x	u_y	u_x	u_y
1	0.062 46	0	0.062 50	0
2	0.057 70	−0.071 73	0.057 73	−0.071 75
3	0.044 16	−0.132 54	0.044 21	−0.132 62
4	0.023 85	−0.173 20	0.023 89	−0.173 18
5	0	−0.187 42	0	−0.187 50

Boundary element stresses

BE no.	Computed			Analytical		
	σ_{xx}	σ_{yy}	σ_{xy}	σ_{xx}	σ_{yy}	σ_{xy}
	0.1078	2.7385	−0.5434	0.1125	2.7290	−0.5542
I	0.3557	2.0545	−0.8583	0.3537	2.0592	−0.8526
	0.5444	1.2222	−0.8128	0.5452	1.2354	−0.8221
	0.1631	0.0787	−0.1092	0.1549	0.0645	−0.1002
II	−0.3549	−0.0768	0.1471	−0.3527	−0.0601	0.1455
	−0.8123	−0.0252	0.1612	−0.8083	−0.0333	0.1641

points on the element corresponding to the Gauss points of the three-point quadrature. (These were chosen as the reference points to allow direct comparison with a linked b.e.–f.e. scheme, discussed later.) It is notable that, even with the coarse discretization of the boundary, there is good correspondence between the two solutions.

In assessing the performance of the linear isoparametric formulation, it is found that the boundary discretization must be significantly finer than for the quadratic scheme to achieve comparable precision in the analysis.

5.3   Boundary element method for a non-homogeneous medium and dominant discontinuities

5.3.1   Continuity conditions on internal surfaces

Although the homogeneous, isotropic continuous model of a rock mass is valid for many design exercises, the capacity to analyse heterogeneous media with continuous planes of weakness extends the range of application of the method. The nature of the problem is illustrated in Figure 5.6. The following discussion of the development of the related boundary element scheme is based on that of Crotty and Wardle (1985). Related procedures have been described by Crouch (1976) and Austin *et al.* (1982).

The conditions to be satisfied for the problem represented in Figure 5.6

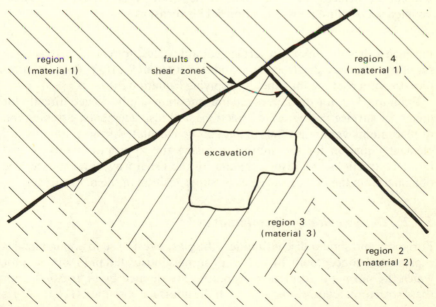

Figure 5.6   Problem definition for b.e. analysis of a heterogeneous medium and continuous planes of weakness.

are the governing equations for elastostatics in each subregion, continuity of traction and displacement at the interfaces between subregions, and the criteria defining slip and separation on the joint surfaces. The procedure followed is to construct a global set of equations, within which all the local conditions are satisfied.

For each homogeneous subregion, S, of the medium, Equation 5.2 must be satisfied. A boundary constraint equation for the subregion can be developed by taking point P, in turn, as each point on the excavation boundary or on the interfaces with adjacent subregions. In developing the discretization of Equation 5.4 for each subregion, it is valid to express the equation in terms of the normal and tangential components of traction and displacements at the nodes, t_n, t_t, u_n, u_t, respectively. Equation 5.12 for each subregion is written

$$\mathbf{T}(S)\mathbf{u}(S) = \mathbf{U}(S)\mathbf{t}(S) \tag{5.14}$$

If no slip and separation occurs at the interface, the conditions for force equilibrium and displacement continuity between subregions S_1 and S_2 require

$$\left.\begin{aligned} t_\alpha(S_1) &= -t_\alpha(S_2) \\ u_\alpha(S_1) &= u_\alpha(S_2) \end{aligned}\right\} \quad \alpha = n, t \tag{5.15}$$

In constructing the complete set of equations for interfaced subregions and excavations within them, explicit boundary conditions are needed for excavation surfaces only. Here half of the set of $(\mathbf{u}, \mathbf{t})$ is specified. For the coupled set of subregions, Equation 5.13 is recast as

$$\mathbf{Ax} = \mathbf{v} \tag{5.16}$$

in which $\mathbf{x}$ lists the unknown surface variables plus, for each interface, one set of the interface displacements and tractions, $\mathbf{u}(S)$, $\mathbf{t}(S)$. For any interface, the other set is eliminated by applying Equation 5.15.

Equation 5.16 for the heterogeneous medium can be solved directly for unknown surface and interface tractions and displacements. Stresses and displacements within any subregion can be calculated from the boundary element equation for the subregion, using the appropriate surface and interface tractions and displacements. In the usual way, total stresses are obtained from the induced stresses by superposition of the field stresses.

5.3.2   Deformation of discontinuities

For joints and similar discontinuities, the elastic response to applied load is represented conceptually by the set of springs illustrated in Figure 5.7a for which the normal deformation properties are defined by Figure 5.7b. For these structural features, the second of the interface conditions (Eqn 5.15) is rewritten as

$$t_\alpha(rs) = k_\alpha \Delta u_\alpha(rs) \quad \alpha = n, t \tag{5.17}$$

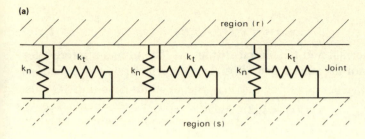

(a)

region (r)

k_n k_t k_n k_t k_n k_t Joint

region (s)

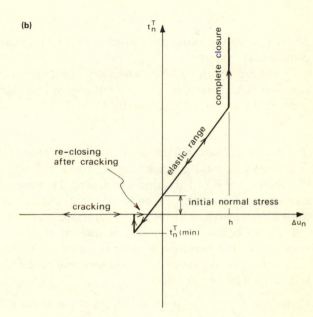

(b)

t_n^T

complete closure

elastic range

re-closing
after cracking

cracking

initial normal stress

t_n^T (min)

h

Δu_n

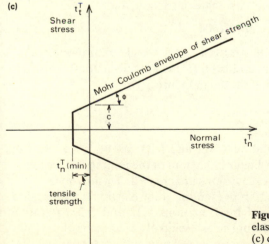

(c)

t_t^T

Shear
stress

Mohr Coulomb envelope of shear strength

ϕ

c

t_n^T (min)

Normal
stress

t_n^T

tensile
strength

Figure 5.7   (a) and (b) Representation of
elastic deformation at a discontinuity;
(c) discontinuity strength properties.

where

$$\Delta u_\alpha(rs) = u_\alpha(s) - u_\alpha(r) \qquad (5.18)$$

and k_n, k_t are normal and tangential stiffnesses. The ultimate interest for joints is in total normal and tangential stresses, for which relevant equations are

$$t_\alpha(r) = k_\alpha \Delta u_\alpha(rs) + t_{\alpha 0}(r) \qquad (5.19)$$

and

$$t_\alpha^T(r) = t_\alpha(r) + t_\alpha^P \qquad (5.20)$$

where $t_{\alpha 0}(r)$ is the initial joint stress prior to joint elastic deformation and t_α^P is the initial field stress.

In setting up the **global boundary constraint equations**, the **stiffness equations** are assembled with the subregion boundary element equations, yielding the system represented by

$$\mathbf{Bx} = \mathbf{v} \qquad (5.21)$$

The coefficient matrix **B** contains blocks of $\mathbf{T}(r)$ and $\mathbf{U}(r)$ for each subregion and blocks of joint stiffness equations. For joint nodes, Equation 5.15 eliminates one set of traction unknowns. However, Equation 5.18, when included in Equation 5.17, indicates that separate displacement variables must be retained for the subregion interfaces defining joints.

Inelastic behaviour of joints is due to the limited tensile and shear strengths of the features, leading to separation and slip under particular states of stress. A generalized **joint strength model**, defined by Figure 5.7c, is described by

$$|t_t^T| \le |t_t^T|(\text{max}) \le c + t_n^T \tan \phi \qquad (5.22)$$

where c is the joint cohesion and ϕ its angle of friction, and

$$t_n^T \ge t_n^T(\text{min}) \qquad (5.23)$$

where $t_n^T(\text{min})$ is the joint tensile strength. If the tensile failure criterion is satisfied, joint separation occurs and both t_n^T and t_t^T vanish. Joint closure is specified by the limiting condition shown in Figure 5.7b.

5.3.3   Solution procedure

The conditions defined in Equations 5.22 and 5.23 can be introduced in Equation 5.21 to model the non-linear behaviour of the jointed system in the state of stress that develops as excavations are created. This involves, first, direct solution of Equation 5.21 representing joint elastic and subregion behaviour, and then, if required by Equations 5.22 and 5.23, iterative solution to determine the equilibrium stresses and displacements at the joint nodes. At each joint node, tests are conducted sequentially for satisfaction of the joint separation condition, the yield condition, and for maximum closure of the joint.

In solving the global boundary constraint equation, the development of slip, separation, and joint closure can lead to instability if care is not taken to impose a realistic stress path. In practice, this requires that the load vector **v** of Equation 5.21 must be established incrementally in the non-linear part of the development of the final state of stress.

Illustration of the performance of the code BITEMJ described by Crotty (1983), which implements these non-linear joint features, is provided in Figure 5.8. When the angle of friction ϕ of the joint is less than the angle of inclination to the joint intersection with the boundary of the circular hole, slip occurs on the plane of weakness. Elementary static analysis (Brady & Brown 1985) indicates that, near the intersection, the tangential component of boundary stress, and therefore all boundary stresses, must be zero. This is confirmed by the b.e. analysis. The variation of normal and shear stress components along the plane of weakness shows the modification of the

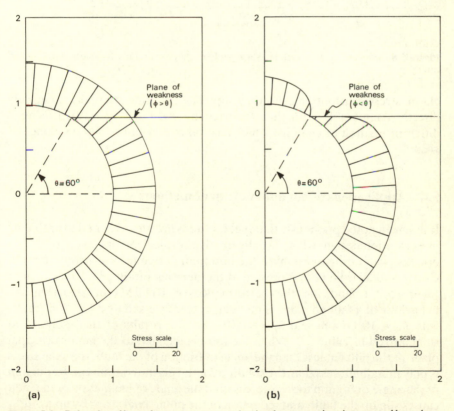

Figure 5.8   Polar plot of boundary stresses around a circular excavation, intersected by a plane of weakness, for hydrostatic field stresses: (a) no slip; (b) with slip ($\phi = 16.3°$);

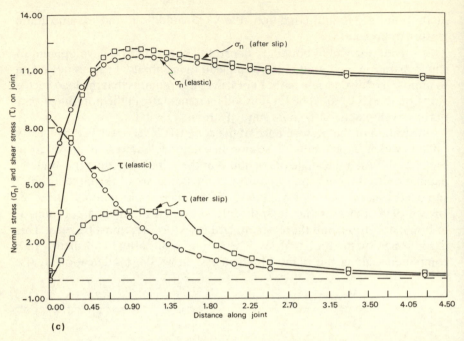

Figure 5.8 *continued*   (c) normal and shear stresses on discontinuity for elastic solution and after slip.

elastic stress distribution caused by slip. The resulting stress distribution is directly comparable with that reported by Austin *et al.* (1982) using a different method of analysis, which also takes account of slip on planes of weakness.

5.4   Applications of boundary element methods

It has been noted previously that b.e. methods are ideally suited to problems in excavation design, which usually involves evaluation of a range of design options or iterating to establish the optimum design for a structure in rock. Crotty and Wardle (1985) presented the demonstration problem shown in Figure 5.9 to indicate the performance of BITEMJ in analysing the behaviour of a faulted pillar. The pillar is transgressed by a 1 m thick fault, with $k_n = 10$ GPa/m and $k_t = 1$ GPa/m. The results of the analysis are summarized in Table 5.2. When the fault is inclined to the major principal plane of the pillar at less than the angle of friction of the fault, the axial stress is only marginally less than the solution for the elastic continuum. Reduction of the angle of friction to a value less than the angle of inclination of the fault causes slip on the fault, and reduction of the pillar axial stress by up to 40%, for the particular conditions applying in this demonstration case.

Table 5.2 Results of numerical analyses of faulted-pillar problem (Fig. 5.9 refers).

	No fault Homogeneous 1 region A	Fault modelled Homogeneous 2 regions B	Homogeneous 4 regions C	Hard pillar 4 regions D	Soft pillar 4 regions E
Elastic					
t_n^T	17.5	17.3	17.0	18.2	14.8
t_t^T	10.1	9.3	9.5	10.2	8.4
Δu_n		−0.5	−0.6	−0.4	−0.8
Δu_t		13.7	13.9	14.5	12.8
σ_p maximum	23.3	22.3	22.3	23.8	19.5
minimum	0.0	0.1	−0.2	−0.1	−0.2
angle	90.0	88.8	89.2	89.1	89.5
Non-linear ($\phi = 25$)					
t_n^T		14.6	13.7	14.9	11.7
t_t^T		6.9	6.5	7.1	5.6
Δu_n		−0.8	−0.9	−0.8	−1.1
Δu_t		75.4	90.3	80.4	118.3
σ_p maximum		17.6	16.4	17.8	14.0
minimum		0.2	−0.1	−0.1	−0.1
angle		84.5	83.6	83.5	83.7

Source: Crotty and Wardle (1985).
Note: Stresses are in MPa, displacements in mm, angles in degrees.

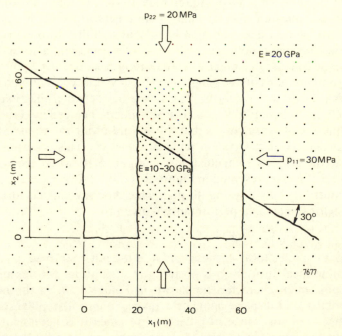

Figure 5.9 Problem geometry for demonstration of b.e. analysis of faulted, non-homogeneous rock mass.

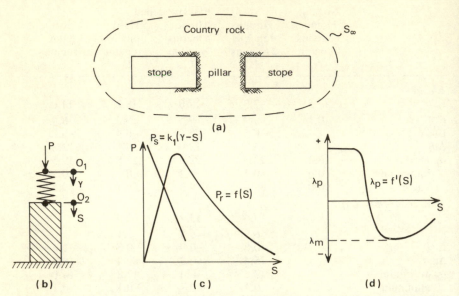

Figure 5.10   Analogue for pillar performance, and specification of stiffness used in stability analysis.

In addition to the routine application of b.e. methods for analysis of stress and displacement in jointed or faulted rock masses, Brady and Brown (1981) describe an application in determining **mine local stiffness**. This is part of the broader problem of determining **mine global stability**. In pillar-supported mine structures, mine global instability involves the potential for sudden and catastrophic collapse of a pillar when its strength is exceeded.

In current methods of assessing mine global stability, a stability index for a pillar may be established by drawing an analogy between pillar loading and a uniaxial compression test on a rock specimen. The general criterion for global stability of a structure is that the second-order variation of the total potential energy of the system must be positive. This criterion may be applied to the simple system illustrated in Figure 5.10, where the mine pillar in Figure 5.10a is represented in Figure 5.10b by a rock specimen, and the mine country rock by a spring through which the specimen is loaded. The condition for stable pillar performance is given by

$$\text{stability index} = k_l + \lambda_p > 0 \qquad (5.24)$$

where k_l is the mine local stiffness, and λ_p is the **pillar stiffness**.

The mine local stiffness k_l is positive, by definition, for the mine local load-displacement characteristic shown in Figure 5.10. In the post-peak domain of the load-displacement performance of the pillar, pillar stiffness λ_p is negative. For this range of pillar performance it is necessary that the stability index defined in Equation 5.24 be evaluated to predict the possibility of maintaining pillar stability.

Mine local stiffness can be estimated by applying distributed loads at the pillar position. The displacements at the pillar location, as the pillar load is relaxed, can be readily calculated using a b.e. code. Pillar stiffness in the post-peak range may be estimated from the pillar elastic stiffness, the nature of the elastic–post-peak relation for the rock material, and the potential for slip on suitably inclined features in the body of the pillar. The code BITEMJ should be particularly useful for analysis of the stability of pillars containing planes of weakness.

5.5   A linked boundary element–distinct element scheme

5.5.1   Analytical formulation

In terms of the conceptual models of rock mass response illustrated in Figure 5.1, that represented by Figure 5.1c is exceptional in the nature of the displacements of near-field rock. Although the displacements for the other three models are of elastic orders of magnitude, a block-jointed medium may experience relatively large block translations and rotations not amenable to analysis by conventional small-strain theory.

The distinct element method originally reported by Cundall (1971) and Cundall et al. (1978) is ideally suited to the analysis of this type of problem. The particular property of the method, of engineering interest, is that it may simulate the behaviour of an aggregation of joint-defined blocks in a manner conceptually and physically consistent with the observed performance of jointed rock masses. By representing the rock which will constitute the near field of an excavation with distinct elements, and the far field with boundary elements, the conceptual advantages of the d.e. method are retained, and rigour in relation to far-field boundary conditions is preserved.

A schematic representation of the nature of the excavation design problem for jointed rock is shown in Figure 5.11. Prior to excavation, the jointed rock

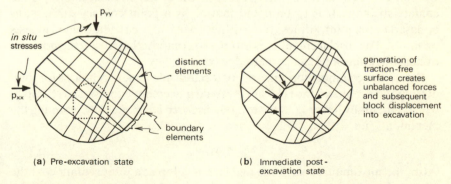

(a) Pre-excavation state

(b) Immediate post-excavation state

Figure 5.11 The response of jointed rock to excavation, modelled with a linked d.e.–b.e. scheme.

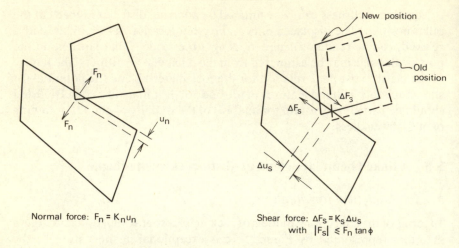

Figure 5.12   Normal and shear interactions of blocks for d.e. analysis.

is in equilibrium with the surrounding rock mass. Generation of the surface of the excavation creates out-of-balance forces at the boundary, and each rock block displaces against either the resistance of its neighbours, or that plus the resistance of any installed support.

For the sake of completeness of the development of the linked b.e.–d.e. scheme, the basis of the distinct element method is summarized as follows. The interaction between two adjacent blocks is defined in the manner shown in Figure 5.12. The normal force F_n developed by block corner-to-edge contact is described by a linear force-displacement law, i.e. for compressive interaction

$$F_n = K_n u_n \qquad (5.25)$$

where u_n is the relative normal displacement of the contact, and K_n is the contact stiffness. It is to be noted that K_n is a point contact stiffness, as opposed to the joint stiffness (i.e. stiffness per unit area) determined in tests on joints. The displacement u_n is estimated computationally by the notional overlap that develops at a corner-edge contact.

Due to the common non-linearity of force-displacement relations for joint deformation in shear, the relevant force-displacement expression is taken to be incrementally linear, and the total shear force is obtained by summing the increments, i.e.

$$\Delta F_s = K_s \Delta u_s \qquad (5.26)$$

Also, the maximum shear force that can develop at a joint is limited by the frictional resistance of the surface. The simplest relation that may be employed is

$$|F_s| \leq F_n \tan \phi \qquad F_n > 0 \tag{5.27}$$

where ϕ is the angle of friction for the surface.

Of course, other expressions defining the shear resistance of surfaces can be readily introduced. Equations 5.25–5.27 describe the simplest conceivable model for interaction between blocks. The explicit formulation of the d.e. method allows more complex and realistic behaviour, such as dilation due to joint roughness, or peak-residual behaviour, to be incorporated readily.

Suppose the net force components and moment acting on a block, arising from its contacts with its neighbours, are F_x, F_y, M. For a block of mass m, the law of motion is

$$\frac{\partial \dot{x}}{\partial t} = \frac{F}{m} \tag{5.28}$$

For a central difference, finite difference scheme defined by

$$\frac{\partial \dot{x}}{\partial t} = \frac{\dot{x}^{(t+\Delta t/2)} - \dot{x}^{(t-\Delta t/2)}}{\Delta t} \tag{5.29}$$

Equation 5.28 becomes, on rearrangement,

$$\dot{x}^{(t+\Delta t/2)} = \dot{x}^{(t-\Delta t/2)} + \frac{F^{(t)}}{m} \Delta t \tag{5.30}$$

For blocks in a gravitational field, the difference equations of motion become

$$\dot{x}_i^{(t+\Delta t/2)} = \dot{x}_i^{(t-\Delta t/2)} + \left(\frac{F_i^{(t)}}{m} + g_i \right) \Delta t \qquad i = 1, 2 \tag{5.31}$$

$$\dot{\theta}^{(t+\Delta t/2)} = \dot{\theta}^{(t-\Delta t/2)} + \frac{M}{I} \Delta t \tag{5.32}$$

where I is the moment of inertia of the block. Block translation and rotation are defined by

$$x_i^{(t+\Delta t)} = x_i^{(t)} + \dot{x}_i^{(t+\Delta t/2)} \Delta t \qquad i = 1, 2$$
$$\theta^{(t+\Delta t)} = \theta^{(t)} + \dot{\theta}^{(t+\Delta t/2)} \Delta t \tag{5.33}$$

where x_i are coordinates of the block centroid, and θ is block rotation about the centroid.

For practical application, viscous damping is introduced to dissipate the kinetic energy and to achieve a stable state. The equation of motion becomes

$$\frac{\partial \dot{x}}{\partial t} = \frac{F}{m} - \alpha \dot{x} + g \tag{5.34}$$

where α is the damping constant for mass proportional damping. Equation 5.31 then becomes

$$\dot{x}_i^{(t+\Delta t/2)} = \left[\dot{x}_i^{(t-\Delta t/2)} \left(1 - \frac{\alpha \Delta t}{2} \right) + \left(\frac{F_i^{(t)}}{m} + g_i \right) \Delta t \right] \bigg/ (1 + \alpha \Delta t/2) \qquad (5.35)$$

with a similar equation for damping of rotation.

In the computational implementation of Equations 5.25–5.35, as described by Cundall in Chapter 4, the damped motion of each block is followed through a series of time steps. Lorig (1984) has shown that convergence to a stable static solution occurs provided the system is over-damped. At the end of the computation, centroidal translations and rotation, nodal displacements and nodal forces, are known for all elements. Element average stresses can be calculated from the expression

$$\bar{\sigma}_{ij} = \frac{1}{v} \sum x_i F_j \qquad (5.36)$$

5.5.2   Linkage algorithm

Linkage of b.e. and d.e. schemes is performed by satisfying the conditions of displacement continuity and force or traction equilibrium at the interface between the two solution domains. In considering the infinite elastic domain interacting with the set of distinct elements, shown in Figure 5.13, the most convenient procedure is to develop a nodal force-displacement relation for the many-noded surface represented by Figure 5.13c. Treated in this way, the surface S represents a many-noded elastic superelement, in the finite element sense.

It was shown earlier that, for both linear and quadratic isoparametric b.e. formulations, induced nodal tractions $\mathbf{t}$ and displacements $\mathbf{u}$ are related through the boundary constraint equation (i.e. Eqn 5.12):

$$\mathbf{Tu} = \mathbf{Ut}$$

This may be recast as

$$\mathbf{U}^{-1}\mathbf{Tu} = \mathbf{t} \qquad (5.37)$$

or

$$\{\mathbf{CU}^{-1}\mathbf{T}\}\mathbf{u} \equiv \mathbf{K}^b\mathbf{u} = \mathbf{Ct} \equiv \mathbf{f}^b \qquad (5.38)$$

In Equation 5.38, $\mathbf{C}$ is the matrix for conversion of distributed tractions to statically equivalent nodal forces, $\mathbf{K}^b$ the stiffness matrix for the many-noded elastic superelement representing the b.e. domain, and $\mathbf{f}^b$ the nodal force vector for the superelement. The matrix $\mathbf{C}$ is determined in the following way. Consider the quadratic, isoparametric boundary element illustrated in Figure 5.4 and defined by Equations 5.8–5.11, with nodes ($i = 1, 2, 3$) and tractions t_j ($j = 1, 2$). Applying a unit virtual displacement at node i only, by the principle of virtual work one obtains

$$f_j^i = \int_{S_k} t_j \bar{N}^i(\xi) \, dS_k \qquad (5.39)$$

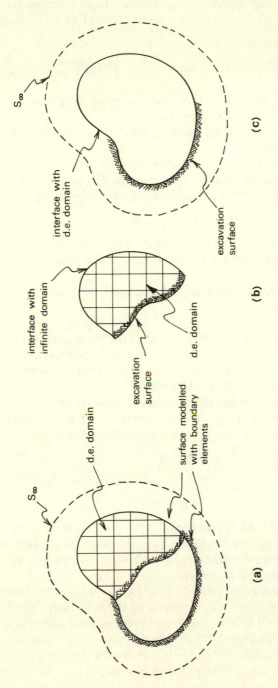

interface with
d.e. domain

excavation
surface

(c)

interface with
infinite domain

excavation
surface

d.e. domain

(b)

S_∞

d.e. domain

surface modelled
with boundary
elements

(a)

S_∞

Figure 5.13 Resolution of a coupled d.e.–b.e. problem into component problems.

where f_j^i is the equivalent nodal point force in the jth coordinate direction at node i, and $\bar{N}^i$ is the appropriate quadratic shape function for the node. Taking account of the linear variation of tractions (Eqn 5.11), Equation 5.39 yields

$$f_j^i = \int_{-1}^{0} [t_j^p N^p(\alpha)] \bar{N}^i(\xi) |J_k(\xi)| \frac{d\xi}{_\alpha d\alpha}\, d\alpha + \int_{0}^{1} [t_j^{p+1} N^p(\beta)] \bar{N}^i(\xi) |J_k(\xi)| \frac{d\xi}{_\beta d\beta}\, d\beta \quad (5.40)$$

where N^p are the linear shape functions. By evaluating these integrals over all boundary elements $(k = 1, n)$, the matrix $\mathbf{C}$ is obtained. It is sparsely populated, and its coefficients are integrals of the products of the linear and quadratic shape functions.

Equation 5.38 describes the relation between a set of internal, induced nodal forces and induced nodal displacements. The induced nodal forces are related to the final, equilibrium nodal forces $\mathbf{q}^e$ by the expression

$$\mathbf{f}^b = \mathbf{q}^e - \mathbf{q}^0 \quad (5.41)$$

where $\mathbf{q}^0$ represents a set of initial forces which can be calculated directly from Equation 5.40, with tractions t_j now those due to the initial field stresses.

Equation 5.38 can be recast, using Equation 5.41 to yield

$$\mathbf{q}^e = \mathbf{K}^b \mathbf{u} + \mathbf{q}^0 \quad (5.42)$$

In this expression, $\mathbf{K}^b$ is the stiffness matrix for the elastic superelement.

Linkage of d.e. and b.e. schemes is achieved by introducing, in Equation 5.42, the displacements of the nodes of the distinct elements which are in contact with the b.e. interface. This yields, directly, the force components $\mathbf{q}^e$ induced at the nodes by the imposed nodal displacements. To satisfy force equilibrium at the interface, the force components $\mathbf{q}^d$ exerted by the b.e. domain on the nodes of the abutting distinct element are given by

$$\mathbf{q}^e + \mathbf{q}^d = 0$$

or

$$\mathbf{q}^d = -\mathbf{q}^e \quad (5.43)$$

Thus, as the d.e. solution proceeds, nodal forces arising from interaction with the b.e. domain are calculated by application of Equations 5.42 and 5.43, and updated at each time step. They become an input for the next cycle of computation for the d.e. domain.

In the linked algorithm described by Lorig and Brady (1984), a linear isoparametric formulation is employed in the b.e. code, and the distinct elements are defined by linear edges. Displacement continuity is therefore assured along the interface between the d.e. and b.e. domains.

One problem which must be resolved in linking the two schemes is that of rigid-body displacements. The problem is analogous to that in f.e. analysis, where the global stiffness matrix is singular unless suitably conditioned by taking account of fixity conditions at particular nodes. The difficulty is expressed in linked b.e.–d.e. analysis when distinct elements are subject to a gravitational body force component, and there is an equivalent stress gradient in the b.e. domain. For the b.e. surface, the set of nodal forces is no longer self-equilibrating in the vertical direction. The effect in the analysis is an upward rigid-body displacement of the complete b.e. domain, which must be removed for satisfactory solution in the complete field. This is achieved by modifying the stiffness matrix, using a reference point in the b.e. domain which has zero displacement for every set of interface displacements. The procedure described by Lemos (1983) is successful in overcoming this problem. It is analogous to the method employed by Banerjee and Butterfield (1977) in indirect b.e. schemes for loaded half-space problems.

The procedure is started by giving the ith entry of the interface displacement vector a unit displacement and setting all other entries equal to zero. The corresponding vector of interface forces for this unit displacement is the ith column of $\mathbf{K}^b$, denoted $\mathbf{f}^i$. The displacements at the reference point P corresponding to the prescribed interface displacement are u_{xp}^{oi}, u_{yp}^{oi}.

In the next step, each of the interface nodes is given a unit displacement in the x direction, resulting in displacements u_{xp}^{rx} and u_{yp}^{rx} at P. The resulting interface forces $\mathbf{f}^{rx}$ are obtained by summing the elements of the odd-numbered columns in each row of $\mathbf{K}^b$. Similarly, unit y-directed displacements at each node result in a nodal force vector $\mathbf{f}^{ry}$ and displacements u_{xp}^{ry} and u_{yp}^{ry} at P.

The requirement is to determine a set of rigid-body displacements which, taken together with the unbalanced nodal forces corresponding to the unit displacement for entry i, results in no relative displacement between the node corresponding to entry i and the point P. Suppose the required displacements are a_x^i, a_y^i. Superimposing on the ith configuration the displacement field due to the nodal displacements a_x^i and a_y^i produces displacements at P given by

$$u_{xp}^i = u_{xp}^{oi} + a_x^i u_{xp}^{rx} + a_y^i u_{xp}^{ry}$$
$$u_{yp}^i = u_{yp}^{oi} + a_x^i u_{yp}^{rx} + a_y^i u_{yp}^{ry} \qquad (5.44)$$

The condition that P does not displace relative to the boundary nodes is

$$u_{xp}^i - a_x^i = 0$$
$$u_{yp}^i - a_y^i = 0 \qquad (5.45)$$

Equations 5.44 and 5.45 can be combined and recast to give

$$u_{xp}^{oi} + a_x^i(u_{xp}^{rx} - 1) + a_y^i u_{xp}^{ry} = 0$$
$$u_{yp}^{oi} + a_x^i u_{yp}^{rx} + a_y^i(u_{yp}^{ry} - 1) = 0 \qquad (5.46)$$

Equations 5.46 can be solved directly for a_x^i, a_y^i, which are to be super-imposed on the ith configuration. The interface force vector becomes

$$\mathbf{f}'^i = \mathbf{f}^i + a_x^i \mathbf{f}^{rx} + a_y^i \mathbf{f}^{ry} \tag{5.47}$$

The vector $\mathbf{f}'^i$ is the ith column of a modified stiffness matrix which ensures that an out-of-balance force equivalent to the ith entry produces no rigid-body displacement of the boundary relative to the reference point.

The procedure is repeated for all columns of the stiffness matrix $\mathbf{K}^b$, resulting in a modified matrix $\mathbf{K}'^b$. This matrix relates nodal forces and displacements subject to the condition that out-of-balance nodal forces produce no rigid-body displacement of the interface surface relative to a point in the medium.

For a given set of induced nodal forces at the interface, the difference in nodal displacements obtained using $\mathbf{K}'^b$ instead of $\mathbf{K}^b$ represents the rigid-body displacement a_x, a_y of the problem domain. These are given by

$$a_x = \sum_{i=1}^{2n} a_x^i u_i$$

$$a_y = \sum_{i=1}^{2n} a_y^i u_i \tag{5.48}$$

where $u_i (=\mathbf{u})$ are the induced nodal displacements. These are used in determining the final displacements at a point in the interior of the b.e. domain, by addition to the displacements induced by $\mathbf{q}^e$ and $\mathbf{u}$ at the point.

5.5.3   Demonstration problems for linked scheme

Verification of the performance of the linked b.e.–d.e. scheme is limited by the absence of closed form solutions for blocky media. Two problems serve to demonstrate the adequacy of the analytical procedure.

The first case is a circular hole in a hydrostatic stress field, reported by Lorig (1984), and illustrated in Figure 5.14a. Square distinct elements were used to model the near-field rock around the excavation, and the normal and shear contact stiffnesses, which confer the elasticity of the system, were assigned equal values. The stress distribution around the excavation is shown in Figure 5.14. Bearing in mind the relative coarseness of the nominally elastic distinct element model compared with the elastic con-tinuum, the correspondence between the analytical and numerical solutions is regarded as satisfactory. It is notable that the stress distribution varies smoothly across the interface between the different solution domains, indicating that the linkage scheme performs correctly. With regard to the displacements around the excavation, the calculated value is 1.2 mm for $\theta = 0, 90°, 180°, 270°$, compared with the analytical value (for the particular

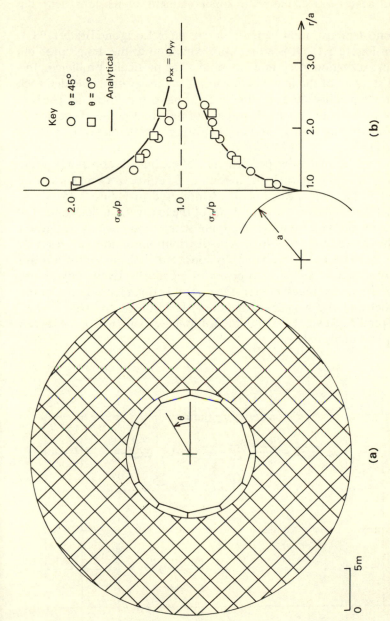

Figure 5.14 Demonstration problem for linked d.e.–b.e. scheme, modelling near-field rock with equivalent jointed medium.

excavation dimensions and elastic properties) of 1.4 mm. The discrepancy is taken to be due to the coarse discretization with rigid elements, and the high contact stiffnesses associated with small element dimensions near the boundary.

The second demonstration exercise, described by Lorig and Brady (1983), is shown in Figure 5.15. It is a first approximation to the design problem posed by an excavation in a jointed medium. The particular problem in this case is the stability of the immediate roof of the excavation. Bedded rock masses of the type illustrated are effectively transversely isotropic. The b.e. component of the code was modified accordingly by the use of the kernel functions for point loads in a transversely isotropic medium, given by Rizzo and Shippy (1970).

In examining the qualitative performance of the linked code, it was found that the key features of rock behaviour in stratified rock masses were reproduced in the numerical model, as indicated in Figure 5.15. These features included slip over the abutments of the excavation, detachment of the immediate roof bed from the overlying strata, and opening of tension cracks in the centre of the roof span. The distributions of normal stress and shear stress in the roof bed were generally consistent with the Voussoir beam model of roof bed behaviour, proposed originally by Evans (1941) and modified by Beer and Meek (1982). One notable divergence between the Evans model and the results of the numerical study was that the bed separation predicted over the complete roof span by Evans was not observed in the computational analysis.

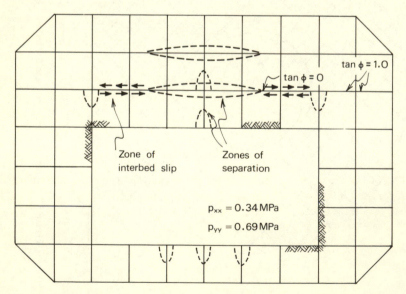

Figure 5.15 Results of d.e.–b.e. analysis of an excavation in stratified rock for the prescribed initial conditions.

5.6   Structural element method for modelling support

Support installed inside an excavation is subject to loading as the rock mass displaces against the resistance developed by the stiffness of the support. A two-dimensional model can show such interaction between rock and installed support, since most of the closure between excavation boundaries occurs after the cross section is excavated during tunnel advance. The development of boundary displacements, and the derived notion of a ground reaction curve, are discussed by Daemen (1977). Useful as this notion is, it is strictly applicable only to axisymmetric problems. The **structural element method**, discussed in detail by Lorig (1984) and Lorig and Brady (1984), is more general in its treatment of rock support mechanics, and rock support interaction.

The essence of the structural element analysis of support systems is the formulation of a stiffness matrix defining the generalized load-displacement behaviour of the support structure. The method forms part of the well established engineering procedures for matrix analysis of linear structural systems. The usual displacement formulation of the method is presented in many texts on structural mechanics, such as that by Ghali and Neville (1978). A structure is resolved into discrete structural elements, assumed to deform in a linearly elastic way under applied load. The stiffness matrix of each element may be established from its simple deformation mechanics, and the stiffness matrix for the structure is constructed by satisfying the conditions for equilibrium and continuity at each node of the structure.

The requirement in the structural element method is to establish the relation between internal loads and element displacements, expressed relative to global reference axes. Figure 5.16a represents a bar element inclined relative to the global axes, which can sustain axial loading only. For nodal forces and displacements expressed relative to the global axes, the force-displacement relation is expressed by

$$\frac{AE}{L}\begin{bmatrix} \lambda^2 & \lambda\mu & -\lambda^2 & -\lambda\mu \\ & \mu^2 & -\lambda\mu & -\mu^2 \\ & & \lambda^2 & \lambda\mu \\ \text{SYM} & & & \mu^2 \end{bmatrix}\begin{bmatrix} u_{x_1} \\ u_{y_1} \\ u_{x_2} \\ u_{y_2} \end{bmatrix} = \begin{bmatrix} f_{x_1} \\ f_{y_1} \\ f_{x_2} \\ f_{y_2} \end{bmatrix} \tag{5.49}$$

where $\lambda = \cos\theta$, $\mu = \sin\theta$.

The beam element shown in Figure 5.16b is subject to transverse forces f_y and bending moments M at nodes 1 and 2. If it has a uniform flexural rigidity EI over its length L, the relation between nodal displacements and forces is expressed, relative to the local X, Y axes, by

$$EI\begin{bmatrix} 12/L^3 & -6/L^2 & -12/L^3 & -6/L^2 \\ & 4/L & 6/L^2 & 2/L \\ & & 12/L^3 & 6/L^2 \\ \text{SYM} & & & 4/L \end{bmatrix}\begin{bmatrix} u_{y_1} \\ \theta_1 \\ u_{y_2} \\ \theta_2 \end{bmatrix} = \begin{bmatrix} f_{y_1} \\ M_1 \\ f_{y_2} \\ M_2 \end{bmatrix} \tag{5.50}$$

When the problem is expressed relative to the global X–Y axes, Equation 5.50 becomes

$$EI \begin{bmatrix} \dfrac{12}{L^3}\mu^2 & -\dfrac{12}{L^3}\lambda\mu & \dfrac{6}{L^2}\mu & -\dfrac{12}{L^3}\mu^2 & \dfrac{12}{L^3}\lambda\mu & \dfrac{6}{L^2}\mu \\[2mm] & \dfrac{12}{L^3}\lambda^2 & -\dfrac{6}{L^2}\lambda & \dfrac{12}{L^3}\lambda\mu & -\dfrac{12}{L^3}\lambda^2 & -\dfrac{6}{L^2}\lambda \\[2mm] & & \dfrac{4}{L} & -\dfrac{6}{L^2}\mu & \dfrac{6}{L^2}\lambda & \dfrac{2}{L} \\[2mm] & & & \dfrac{12}{L^3}\mu^2 & -\dfrac{12}{L^3}\lambda\mu & -\dfrac{6}{L^2}\mu \\[2mm] & \text{SYM} & & & \dfrac{12}{L^3}\lambda^2 & \dfrac{6}{L^2}\lambda \\[2mm] & & & & & \dfrac{4}{L} \end{bmatrix} \begin{bmatrix} u_{x_1} \\[2mm] u_{y_1} \\[2mm] \theta_1 \\[2mm] u_{x_2} \\[2mm] u_{y_2} \\[2mm] \theta_2 \end{bmatrix} = \begin{bmatrix} f_{x_1} \\[2mm] f_{y_1} \\[2mm] M_1 \\[2mm] f_{x_2} \\[2mm] f_{y_2} \\[2mm] M_2 \end{bmatrix} \quad (5.51)$$

For the general case of loading, where a beam is subject to combined axial, transverse and flexural loads, as shown in Figure 5.16c, the stiffness matrix **K** is obtained by the superposition of the separate stiffness matrices in Equations 5.49 and 5.51, i.e.

$$\mathbf{K} = \frac{E}{L} \begin{bmatrix} A\lambda^2 + \dfrac{12I}{L^2}\mu^2 & \left(A - \dfrac{12I}{L^2}\right)\lambda\mu & \dfrac{6I}{L}\mu & -\left(A\lambda^2 + \dfrac{12I}{L^2}\mu^2\right) & -\left(A - \dfrac{12I}{L^2}\right)\lambda\mu & \dfrac{6I}{L}\mu \\[2mm] & A\mu^2 + \dfrac{12I}{L^2}\lambda^2 & -\dfrac{6I}{L}\mu & -\left(A - \dfrac{12I}{L^2}\right)\lambda\mu & -\left(A\mu^2 + \dfrac{12I}{L^2}\lambda^2\right) & -\dfrac{6I}{L}\mu \\[2mm] & & 4I & -\dfrac{6I}{L}\mu & -\dfrac{6I}{L}\lambda & 2I \\[2mm] & & & A\lambda^2 + \dfrac{12I}{L^2}\mu^2 & \left(A - \dfrac{12I}{L^2}\right)\lambda\mu & -\dfrac{6I}{L}\mu \\[2mm] & \text{SYM} & & & A\mu^2 + \dfrac{12I}{L^2}\lambda^2 & \dfrac{6I}{L}\lambda \\[2mm] & & & & & 4I \end{bmatrix}$$

$$(5.52)$$

where **K** relates the global displacements and internal loads defined by

$$\mathbf{u}^T = [u_{x_1} u_{y_1} \theta_1 u_{x_2} u_{y_2} \theta_2] \quad \text{and} \quad \mathbf{f}^T = [f_{x_1} f_{y_1} M_1 f_{x_2} f_{y_2} M_2]$$

In standard structural analysis, the stiffness matrix of the structure is developed by taking account of the connectivity of the nodes. The global equation $\mathbf{K}^s \mathbf{u}_s = \mathbf{r}$ is established by inserting the applied forces in the load vector **r**, and appropriate boundary conditions are accounted for in the displacement vector **u**, and by modifying the global stiffness matrix. After

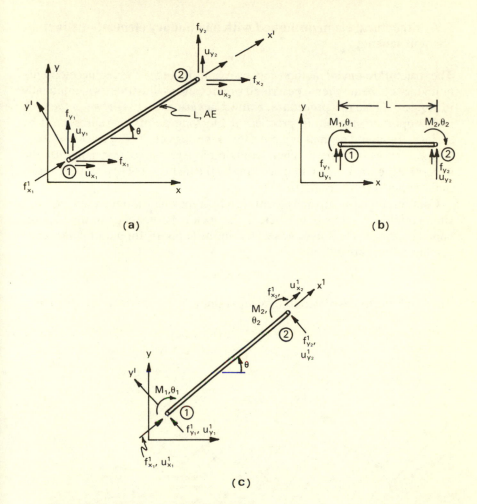

Figure 5.16   Specification of loads and element geometry for a structural element.

solution to determine unknown displacements, the internal loads in any element are determined from the element stiffness matrix and its nodal displacements. These yield the stress components directly from the element geometry.

The problem of interaction between a rock mass and any installed support differs from the typical structural engineering problem in that no external resultant load vector **r** is known. Instead, these forces develop as displacements are imposed on the structure by the adjacent rock. The equilibrium state of loading and deformation is determined by the properties of both the support system and the rock mass interacting with it.

5.7   Structural elements linked with a boundary element–distinct element scheme

The structural element method may be readily linked to the distinct element–boundary element scheme described previously, as illustrated schematically in Figure 5.17. In this procedure, contact at various points between the rock and support, consisting in practice of blocking points, is represented by springs orientated normal and parallel to the surface. The springs are taken to have known stiffness. Their contribution to the performance of the support structure is taken into account by the inclusion of appropriate terms in the support stiffness matrix.

Computationally, linkage is achieved by imposing continuity and equilibrium conditions at the points where the rock mass bears upon the support contact springs. The forces generated in the support, for a particular set of displacements, are given by

$$\mathbf{r}^s = \mathbf{K}^s \mathbf{u} \qquad\qquad (5.53)$$

Therefore the forces $\mathbf{r}^d$ imposed by the support at d.e. contacts are given by

$$\mathbf{r}^d = -\mathbf{r}^s \qquad\qquad (5.54)$$

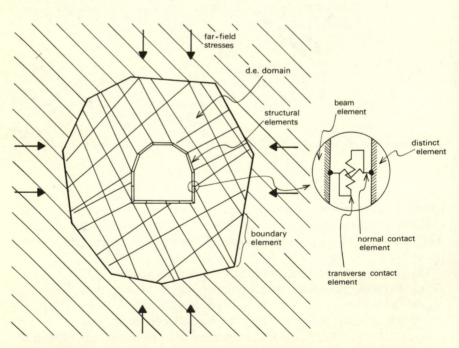

Figure 5.17   Linkage of boundary elements, distinct elements, and structural elements.

These forces are introduced in the next computational cycle for the d.e. domain. Thus, the support forces mobilized by rock displacements can be updated in each computational cycle, in the same way as described previously for the interface forces between the infinite domain and the d.e. domain.

The problem selected to illustrate the performance of the linked s.e.–d.e.–b.e. scheme is shown in Figure 5.18a. It represents a circular hole in a regularly jointed medium, with the problem parameters defined in Table 5.3. A fault, which has a low coefficient of friction, intersects the span of the excavation, defining a wedge in the crown.

In describing the properties of the support system, important dimensionless parameters are the compressibility ratio, C, the flexibility ratio, F, and the contact stiffness ratio, CSR. These are defined by

$$C = E_m R / E_s t$$

$$F = E_m R^3 / 6 E_s I_s$$

$$\text{CSR} = \text{contact transverse stiffness/contact normal stiffness}$$

where

$$E_m, E_s = \text{Young's moduli of rock medium, and support}$$

$$R, t = \text{radius and effective thickness of the support}$$

$$I_s = \text{moment of inertia of the support cross section}$$

In this problem, E_m was taken as the mean of E_1 and E_2, and E_s was taken to be 25 GPa. The contact normal stiffness was 62 GN/m/m.

Table 5.3 Problem parameters for demonstration problem in rock support analysis.

Joint properties

	Normal and shear stiffness (GPa/m)	Coefficients of friction
primary joints	5.82	1.19
secondary joints	3.70	1.19
fault	2.60	0.01

Properties of transversely isotropic equivalent continuum
  $E_1 = 7.39$ GPa
  $E_2 = 10.86$ GPa $\nu_1 = \nu_2 = 0$
  $G_{12} = 4.4$ GPa

Far-field stresses
$p_{xx} = 1.2$ MPa
$p_{yy} = 1.2$ MPa + gravity gradient
$p_{xy} = 0$

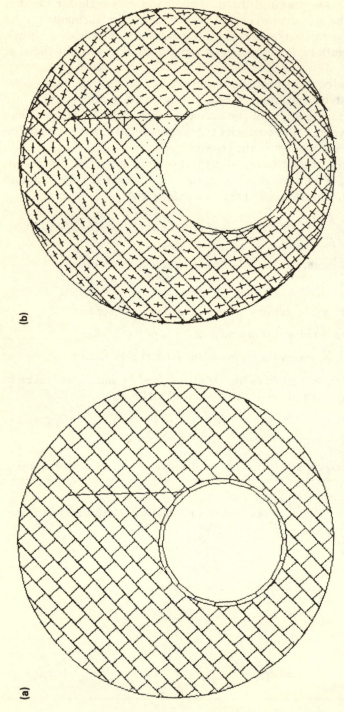

Figure 5.18 Problem geometry for demonstration of linked d.e.–b.e. code performance (a) and tensor plot of principal stresses (b).

Analysis of the problem shown in Figure 5.18a was performed initially without any installed support. This showed that when the coefficient of friction for the fault (μ_f) exceeded 0.09, the wedge in the crown was stable. When μ_f was reduced to 0.05, the entire wedge displaced vertically from the crown, except for the uppermost three blocks. Figure 5.18b shows the stress distribution in the near-field rock for this condition. It illustrates the obvious de-stressing of the wedge in the crown.

The results of analysis of the state of loading in the support, for the case of support installed immediately after excavation and prior to any relaxation in the jointed medium, are shown in Figures 5.19a and b for contact stiffness ratios of 0.0 and 0.2. Figure 5.19a confirms that for no transverse contact stiffness the load vectors are all directed radially, and the thrust is uniform around the support circumference. The effect of the crown wedge is expressed as a significant circumferential variation of bending moment, with high absolute values of the bending moment in the vicinity of the crown wedge. This will produce locally high bending stresses. This behaviour is, of course, completely consistent with the field behaviour of support subject to local loading by an unstable block in the boundary of an excavation.

Comparing the cases where the contact stiffness ratio is 0.2, shown in Figure 5.19b, with that for CSR = 0 (i.e. Fig. 5.19a), it is observed that the properties of blocking between the rock mass and the installed support have a significant effect on load developed in the support. The thrust is no longer uniform, the high local bending movement is considerably reduced, and the load vector is generally inclined to the support surface. A detailed analysis of the states of stress would show that the existence of a transverse contact stiffness generates an improved state of stress in the support.

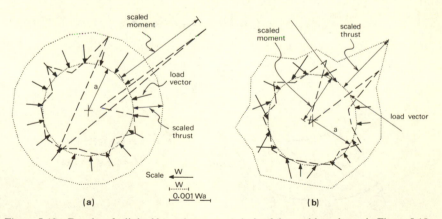

Figure 5.19 Results of a linked b.e.–d.e.–s.e. analysis of the problem shown in Figure 5.18a, for contact stiffness ratios of 0.0 (a), and 0.2 (b). Scale factor W is the weight of the destressed crown wedge implied in Figure 5.18b.

5.8   A linked boundary element–finite element scheme

A linked b.e.–f.e. scheme is the most appropriate numerical procedure for analysis of problems of the type illustrated in Figure 5.1d. A f.e. code modelling elastoplastic or similar constitutive behaviour may be needed to model the near-field randomly jointed or fractured rock around an excavation.

A schematic representation of a linked scheme is as illustrated in Figure 5.13, but with finite elements replacing the distinct elements. Excavations may be bounded by boundary elements or finite elements, on which the specified boundary conditions must be satisfied. At the interface between the f.e. domain and the b.e. domain, the equilibrium and continuity conditions for traction and displacement (Eqn 5.15) must be satisfied.

For the typical finite element shown in Figure 5.20, the well known displacement formulation of the f.e. method (Zienkiewicz 1977) establishes a relation between nodal forces and induced displacements. At any point P in the element, displacements $\mathbf{u}^p$ are related to the nodal displacements $\mathbf{u}^e$ by

$$\mathbf{u}^p = \mathbf{N}\mathbf{u}^e \tag{5.55}$$

where $\mathbf{N}$ is a matrix of interpolation (or shape) functions of the form described previously, except now written as appropriate products for two dimensions. Induced strains $\boldsymbol{\varepsilon}^p$ are related to induced displacements by the expression

$$\boldsymbol{\varepsilon}^p = \mathbf{L}\mathbf{u}^p = \mathbf{L}\mathbf{N}\mathbf{u}^e = \mathbf{B}\mathbf{u}^e \tag{5.56}$$

where

$$\mathbf{L} = \begin{bmatrix} \dfrac{\partial}{\partial x} & 0 \\ 0 & \dfrac{\partial}{\partial y} \\ \dfrac{\partial}{\partial y} & \dfrac{\partial}{\partial x} \end{bmatrix} \tag{5.57}$$

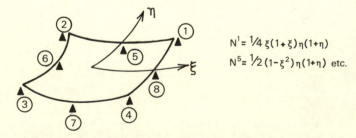

$$N^1 = \tfrac{1}{4}\,\xi(1+\xi)\eta(1+\eta)$$
$$N^5 = \tfrac{1}{2}\,(1-\xi^2)\eta(1+\eta) \text{ etc.}$$

Figure 5.20   Element geometry for a quadratic finite element.

and, for node i,

$$
\mathbf{B}_i = \begin{bmatrix} \dfrac{\partial N^i}{\partial x} & 0 \\[2ex] 0 & \dfrac{\partial N^i}{\partial y} \\[2ex] \dfrac{\partial N^i}{\partial y} & \dfrac{\partial N^i}{\partial x} \end{bmatrix}
\tag{5.58}
$$

The total state of stress within an element is obtained from the induced strains, the elasticity matrix $\mathbf{D}$, and the initial stresses $\boldsymbol{\sigma}_0^p$, i.e.

$$
\boldsymbol{\sigma}^p = \mathbf{D}\boldsymbol{\varepsilon}^p + \boldsymbol{\sigma}_0^p
\tag{5.59}
$$

The f.e. analysis is formulated in terms of the set of nodal (internal) forces, $\mathbf{q}^e$, which is statically equivalent to the forces transmitted between elements at their interface contacts. For no induced body forces, the principle of virtual work yields

$$
\mathbf{q}^e = \int_{V^e} \mathbf{B}^T \boldsymbol{\sigma}^p \, dV
\tag{5.60}
$$

From Equations 5.56 and 5.59, it is found that

$$
\mathbf{q}^e = \mathbf{K}^e \mathbf{u}^e + \mathbf{f}^e
\tag{5.61}
$$

where

$$
\mathbf{K}^e = \int_{V^e} \mathbf{B}^T \mathbf{D} \mathbf{B} \, dV
\tag{5.62}
$$

and

$$
\mathbf{f}^e = \int_{V^e} \mathbf{B} \boldsymbol{\sigma}_0^p \, dV
\tag{5.63}
$$

It is thus a straightforward exercise to determine, using the known element geometry, material properties and field stresses, the element stiffness matrix $\mathbf{K}^e$, and the equivalent external force vector $\mathbf{f}^e$.

In the preceding discussion of linked b.e. and d.e. schemes, Equation 5.42, i.e.

$$
\mathbf{q}^e = \mathbf{K}^b \mathbf{u} + \mathbf{q}^0
$$

was derived, and a technique was established for constructing the stiffness matrix $\mathbf{K}^b$ for the elastic superelement representing the b.e. domain. Equation 5.42 is completely compatible with Equation 5.61 in which $\mathbf{K}^e$ represents the stiffness matrix for the conventional finite elements. Therefore, it is possible to assemble a global force-displacement relation following the usual procedures for the f.e. method, except that, in this case, no attempt is made to exploit any assumed symmetry in the matrix $\mathbf{K}^b$. The procedure described by Hinton and Owen (1977) may be followed. The resultant system of equations is written

$$\mathbf{K}^g \mathbf{u} = \mathbf{r} \tag{5.64}$$

where $\mathbf{K}^g$ is the global stiffness matrix, $\mathbf{u}$ the vector of unknown nodal displacements, and $\mathbf{r}$ the vector of applied nodal loads. $\mathbf{K}^g$ is, in general, sparsely populated and non-symmetric.

After solution of Equation 5.64, the displacement components are known at all nodes of the f.e. and b.e. domains. Application of Equations 5.56 and

Table 5.4 Comparison of coupled finite element–boundary element results with Kirsch's solution (Fig. 5.21 refers).

Induced nodal displacements

Node no.	Computed u_x	u_y	Analytical u_x	u_y
1	0.060 95	0	0.062 50	0
2	0.056 72	−0.070 92	0.057 73	−0.071 75
3	0.044 30	−0.132 68	0.044 21	−0.132 62
4	0.023 05	−0.172 19	0.023 89	−0.173 18
5	0	−0.185 94	0	−0.187 50

Finite element stresses

f.e. no.	Computed σ_{xx}	σ_{yy}	σ_{xy}	Analytical σ_{xx}	σ_{yy}	σ_{xy}
	0.026 45	2.760 6	−0.216 34	0.081 8	2.799 4	−0.207 5
	0.423 50	2.119 4	−0.695 49	0.338 3	2.011 5	−0.751 9
	0.500 78	0.933 2	−0.617 32	0.475 2	0.872 5	−0.618 9
	0.177 67	2.370 8	−0.131 05	0.221 0	2.405 1	−0.147 2
I	0.305 54	1.883 4	−0.521 86	0.291 8	1.876 0	−0.488 2
	0.219 98	0.992 8	−0.484 66	0.261 9	1.039 2	−0.478 4
	0.372 15	2.091 1	−0.076 31	0.299 1	2.120 4	−0.087 4
	0.249 49	1.742 0	−0.387 32	0.255 2	1.766 1	−0.314 0
	0.101 73	1.119 4	−0.387 98	0.121 0	1.141 4	−0.375 9
	0.244 37	0.329 1	−0.321 15	0.301 3	0.350 9	−0.321 6
	−0.437 31	−0.087 5	0.022 39	−0.336 6	−0.013 1	0.075 6
	−0.817 99	0.039 0	0.041 87	−0.858 8	−0.022 4	0.062 3
	0.151 78	0.622 1	−0.330 95	0.111 3	0.587 5	−0.335 4
II	−0.304 37	0.100 2	−0.064 82	−0.292 1	0.124 3	−0.095 1
	−0.561 00	0.000 0	−0.020 37	−0.592 1	−0.033 9	−0.004 1
	0.020 30	0.863 9	−0.316 69	−0.010 2	0.747 8	−0.333 5
	−0.231 84	0.242 5	−0.121 88	−0.255 4	0.234 2	−0.195 6
	−0.378 03	−0.069 8	−0.055 47	−0.410 7	−0.008 8	−0.044 4

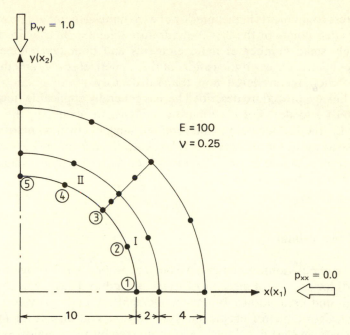

Figure 5.21 Problem geometry and field stresses to demonstrate the performance of a linked f.e.–b.e. scheme.

5.59 allows direct determination of the state of stress at particular points within the element, conveniently the Gauss points of the element.

Determination of stresses and displacements in the infinite domain requires knowledge of nodal tractions for all points defining the inner surface of the elastic superelement. Excavation-induced tractions are calculated by inserting the known nodal displacements in Equation 5.12 and solving directly for **t**. For the case of nodes which lie on the traction-free surface of an excavation, this procedure provides a useful test of the consistency of the induced tractions, since it must return the known values specified initially.

When **t** and **u** are known for the b.e. domain, stresses around the interface, and stresses and displacements at interior points in the domain, are calculated by the usual b.e. procedure.

The problem shown in Figure 5.21 has been used to demonstrate the performance of the quadratic, isoparametric, linked b.e.–f.e. scheme. The choice of the relatively coarse discretization for both the f.e. and b.e. domains and the uniaxial stress field were designed to examine how well the linked algorithm handled the high boundary stress and displacement gradients around the circular hole. The results are tabulated in Table 5.4.

The stress components in the interior of each finite element are provided for the 9 Gauss points of the 3×3 quadrature scheme. Bearing in mind the relatively small number of finite elements and boundary elements, the correspondence between the results of the numerical study, and the closed form solution, is considered more than satisfactory.

The linkage procedure described here is generally applicable for coupling b.e. and f.e. codes. For elastoplastic analysis, for example, the stiffness matrix for the b.e. domain is the required tangent stiffness matrix for the iterative analysis. Once the matrix $\mathbf{K}^b$ is determined from the b.e. scheme, it is therefore a simple exercise to link a b.e. code to a finite element code proven for elastoplastic analysis.

5.9 Conclusions

Several computational schemes have been described which provide efficient methods for analysis of the state of stress and induced displacement around underground excavations. Boundary element methods provide the most efficient algorithms for elastic analysis of stress and displacement, or for cases where an elastic rock mass is transgressed by a few major continuous planes of weakness. For other cases, linked boundary element–distinct element or linked boundary element–finite element schemes are both computationally efficient and conceptually appropriate methods of analysis, particularly in relation to the complex behaviour of the near-field rock around excavations. The advantage of the linked b.e.–d.e. scheme is that it can be easily linked to a structural element method of analysis. This provides a powerful method for detailed analysis of the interaction between rock around an excavation and support installed within the opening.

References

Austin, M. W., J. W. Bray and A. M. Crawford 1982. A comparison of two indirect boundary element formulations incorporating planes of weakness. *Int. J. Rock Mech. Min. Sci. Geomech. Abstr.* **19**, 339–44.

Banerjee, P. K. and R. Butterfield 1977. Boundary element methods in geomechanics. In *Finite elements in geomechanics*, G. Gudehus (ed.), ch. 16. London: Wiley.

Beer, G. and J. L. Meek 1982. Design curves for roof and hangingwalls in bedded rock based on Voussoir beam and plate solutions. *Trans. Instn Min. Metall.* **91**, A18–22.

Brady, B. H. G. 1979. A boundary element method of stress analysis for non-homogeneous media and complete plane strain. *Proc. 20th US Symp. Rock Mech.*, Austin, 243–50.

Brady, B. H. G. and E. T. Brown 1981. Energy changes and stability in underground mining: design applications of boundary element methods. *Trans. Instn Min. Metall.* **90**, A61–8.

Brady, B. H. G. and E. T. Brown 1985. *Rock mechanics for underground mining.* London: Allen & Unwin.

Crotty, J. M. 1983. *User's manual for BITEMJ – two-dimensional stress analysis for piecewise homogeneous solids with structural discontinuities.* Melbourne: CSIRO (Australia) Division of Geomechanics.

Crotty, J. M. and L. J. Wardle 1985. Boundary integral analysis of piecewise homogeneous media with structural discontinuities. *Int. J. Rock Mech. Min. Sci. Geomech. Abstr.* **22**, 419–27.

Crouch, S. L. 1976. Solution of plane elasticity problems by the displacement discontinuity method. *Int. J. Num. Meth. Engng* **10**, 301–43.

Cruse, T. A. 1969. Numerical solutions in three-dimensional elastostatics. *Int. J. Solids Struct.* **5**, 1259–74.

Cundall, P. A. 1971. A computer model for simulating progressive large scale movements in blocky rock systems. In *Rock fracture*, Proc. Int. Symp. Rock Fracture, Nancy, Paper II–8.

Cundall, P. A. 1986. Distinct element models of rock and soil structure. In *Analytical and computational methods in engineering rock mechanics*, E. T. Brown (ed.), 129–63. London: Allen & Unwin.

Cundall, P. A., J. Marti, P. T. Beresford, N. C. Last and M. I. Asgian 1978. *Computer modelling of jointed rock masses.* Technical Report N-78-4, US Army Engineers Waterways Experiment Station, Vicksburg, Mississippi.

Daemen, J. J. K. 1977. Problems in tunnel support mechanics. *Underground Space* **1**, 163–72.

Evans, W. H. 1941. The strength of undermined strata. *Trans. Instn Min. Metall.* **50**, 475–523.

Ghali, A. and A. M. Neville 1978. *Structural analysis: A unified classical and matrix approach*, 2nd edn. London: Chapman & Hall.

Hinton, E. and D. R. J. Owen 1977. *Finite element programming.* London: Academic Press.

Lemos, J. S. 1983. *A hybrid distinct element–boundary element method for the half-plane.* MS thesis, University of Minnesota.

Lorig, L. J. 1984. *A hybrid computational model for excavation and support design in jointed media.* PhD thesis, University of Minnesota.

Lorig, L. J. and B. H. G. Brady 1983. An improved procedure for excavation design in stratified rock. In *Rock mechanics – theory – experiment – practice*, Proc. 24th US Symp. Rock Mech., College Station, Texas, 577–85. College Station, Tex.: Association of Engineering Geologists.

Lorig, L. J. and B. H. G. Brady 1984. A hybrid computational scheme for excavation and support design in jointed rock media. In *Design and performance of underground excavations*, Proc. ISRM Symp., Cambridge, E. T. Brown and J. A. Hudson (eds), 105–12. London: British Geotechnical Society.

Rizzo, F. J. and D. J. Shippy 1970. A method for stress determination in plane anisotropic elastic bodies. *J. Comp. Mat.* **4**, 36–61.

Wassyng, A. 1982. Solving $Ax = b$: a method with reduced storage requirements, *SIAM J. Numer. Anal.* **19**, 197–204.

Watson, J. O. 1979. Advanced implementation of the boundary element method for two- and three-dimensional elastostatics. In *Developments in boundary element methods – 1*, P. K. Banerjee and R. Butterfield (eds), 31–63. Barking, Essex: Applied Science Publishers.

Yeung, D. and B. H. G. Brady 1982. A hybrid quadratic isoparametric finite element – boundary element code for underground excavation analysis. In *Issues in rock mechanics*, Proc. 23rd U.S. Symp. Rock Mech., Berkeley, R. E. Goodman and F. E. Heuze (eds), 692–703. New York: Society of Mining Engineers, American Institute of Mining, Metallurgical and Petroleum Engineers.

Zienkiewicz, O. C. 1977. *The finite element method*, 3rd edn. London: McGraw-Hill.

6 Stability analysis of infinite block systems using block theory

GEN HUA SHI and R. E. GOODMAN

6.1 Introduction

An excavation in jointed rock involves construction in an **imperfectly known** environment with an **infinity of failure modes**, ranging from falls of individual blocks to the collapse of large rock volumes. Block theory provides a means of analysing such an excavation, taking into account all potential modes of failure and encompassing the very important attribute of **three-dimensionality** for rock blocks. In this chapter, we describe recent developments that extend applications of block theory to infinite systems of **united blocks**. A united block is a non-convex system of blocks formed by the unions of individual, convex blocks. We discuss the geometry and topology of such systems, their stability analysis, and concepts for determining support requirements. Although an extension of topics discussed in the book *Block theory and its applications to rock engineering* (Goodman & Shi 1985), most of this material has not been previously published.

The methods of block theory developed from practical need in an actual construction project, and although the ideas were formulated with modern mathematics they are not mathematical abstractions. On the contrary, these methods are immediately applicable in actual construction and mining engineering.

The fundamental principle underlying the applications of block theory is that rock failure starts with the movement of a **key block** located on the periphery of an excavation. Workers are forewarned of this movement by observable **joint opening**, which is a geometrical consequence of the shifting of one or more blocks. Such movements can be prevented by **presupport** or by the selection of a **geometric design** that precludes the existence of key blocks. Without key blocks, excavation failure is assumed to be impossible.

6.2   Input for analysis of jointed rock

Because of the important attributes of joints, the **input data** one can hope to acquire about rock for engineering analysis are prioritized differently than in other areas of applied mechanics. **Joints** disrupt the continuity and strength of the rock mass and their role is dominant over that of the strength and deformability of rock material and the state of stress around the excavation. However, the geological survey conducted prior to construction or mining is normally able to provide only incomplete data about the joints, consisting of their orientations and statistical descriptions of their lengths and spacings. Later, during the period of excavation, it is possible to observe the two-dimensional traces of individual joints in the excavated surface, but not their planar extensions inside the rock mass.

In addition to joint data, an important class of input for block theory is the complete description of the **free surfaces** determining the three-dimensional periphery of the excavation. A dam foundation, for example, is formed by the lower rock surface beneath the dam and the two abutments on either side. Underground galleries contain at least six free planes, twelve edges, and eight corners. Tunnels are even more complex as their interior surfaces are curved. Stability analysis has to consider all the free planes and all their real combinations. The most dangerous types of blocks are frequently those that involve several free planes in combination, for example edges and corners of underground chambers, intersections of tunnels, and portals. The portals of tunnels involve the combination of tunnel periphery and free surfaces of the ground.

As in all engineering projects, the **forces** affecting the system need to be evaluated. In rock engineering, all forces except those of self-weight are variable or imperfectly known. These include forces of water, earthquakes, and blasting, and the forces associated with changing states of stress. For design, forces must be considered to change with time and space in both magnitude and direction.

These features of the input quantities make it quite difficult to employ conventional applied mechanics procedures in problems of excavation stability. *The influence of the unknown aspects of jointing make correct stress–strain and displacement computations almost impossible.* Moreover, stress–strain computations do not naturally determine the mode of failure appropriate for the discontinuities and forces, whereas the mode of failure must be known to select the right system of reinforcement. Also, stress–strain computations do not give the support force from a single analysis, as the force required to assure a certain degree of safety must respect all cases involving potential failure. Design based on the many repeated runs necessary to cover all conditions is inconvenient, even when compared with design based on observational or empirical procedures or physical model studies.

6.3 Proposed methodology

The application of block theory accepts the above realities concerning the quality and completeness of input information. It incorporates all essential, reliable data about the joints deduced from the surface, the drill holes, and from logs of previously excavated surfaces as well as friction angles from tests and observations of joints. The object of the analysis of infinite block systems is to compute a smallest safe support pressure for any combination of excavation surfaces and joint sets. The methods avoid extensive numerical operations. The steps in the procedure are as follows.

(1) Divide all the innumerable key blocks into finite **classes** based entirely on the sets of joints comprising them. The basis for this subdivision is the recognition of **joint pyramids**, represented by the symbol JP. (These will be discussed further below.)
(2) For each JP, find a **minimum stable key block** or **maximum unstable key block.** If the mapping of exposed rock surfaces has been able to produce a map of joint traces, it will be possible at this step to delineate the maximum key block.
(3) Determine the required **support force** and **geometric design** to stabilize the maximum unstable key block and all smaller blocks.

A feature of the approach outlined above is that it is based on **logical** rather than **numerical** operations. This makes it attractive to those experienced engineers who operate largely on intuition.

In this chapter we present ideas and logic, reinforced at first by **two-dimensional examples**. The extension of these examples to real three-dimensional problems is straightforward, using either the stereographic projection or mathematics. To show this, a series of three-dimensional examples is given in connection with complete analysis of tunnels and underground chambers.

The joint pyramids are represented in the stereographic projection by circular polygons formed on the surface of a reference sphere circumscribed about the origin, which is the common vertex of all JPs. In two dimensions, the JPs reduce to angles subtended from an origin. A projection is not necessary in two-dimensions as the JP itself can be drawn directly in the geological section. The examples given here, involving only two sets of joints, can easily be generalized to cases with a greater number of joint sets. Although the resulting shapes of united key blocks formed by more than two sets of joints would be more complicated than those depicted, the corresponding JPs are not more difficult to comprehend.

In this chapter our main aim is to communicate fundamental ideas and perceptions, not to present complex formulations or algorithms. The methods have been rigorously established using mathematics; intuitive explanations have been committed to strict mathematical proofs.

6.4   Joint pyramids of united key blocks

Figures 6.1–6.8 present the conditions of a typical dam foundation in hard, jointed rock. Investigations have revealed the orientations of the two joint sets, and holes have been drilled to locate them in position. Since joint sets include an infinity of individuals, there will always be more below the deepest exploration. In these figures, we assume that the joints are sufficiently long to divide the rock mass into blocks through their mutual intersections. The free surface being infinitely large, there are an infinite number of united key blocks and these blocks can be infinitely voluminous. Using the principle of the JP, it will be shown that it is, in fact, possible to introduce a finite number of rock anchors to stabilize this infinity of key blocks.

JP connotes joint pyramid. In two dimensions, a joint pyramid consists of the angle between two adjacent rays on a diagram where joint planes (lines) are drawn for all sets to pass through a single origin. In Figure 6.1b, we have drawn lines 1 and 2 through the origin (0,0) parallel to the joints of sets 1 and 2 respectively. Lines 1 and 2 divide the plane into four angles, denoted 00, 10, 11, and 01. These designations are ordered so that the first number identifies a half-space of the first joint set and the second number identifies a half-space of the second joint set. The digit 0 identifies the upper half-space,

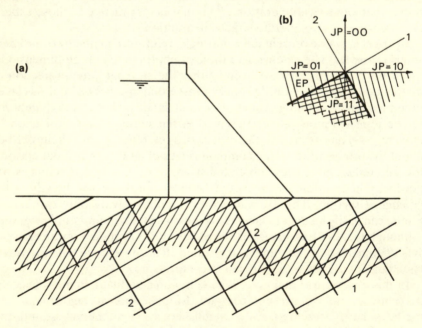

Figure 6.1   (a) The ruled section shows one of the infinite blocks of JP 11; (b) JP 11 has an intersection with EP.

i.e. the half-plane above the joint; similarly, the digit 1 identifies the lower half-space, i.e. the half-plane below the joint. For example, JP 10 is the region simultaneously below line 1 and above line 2.

In the following discussion we shall consider all united key blocks corresponding to each JP.

(a) *JP = 11.* This belongs to any single or united block that is always below the joint segments of both sets, like the ruled area of Figure 6.1a. In this case, the free surface is horizontal and the rock mass is below it. We draw a line parallel to the free surface through the origin (0,0) in Figure 6.1b and designate the half-plane below this line by the term EP for excavation pyramid. In Figure 6.1b, we observe that EP and JP 11 have common rays. (A **ray** is a straight line originating at the origin.) Therefore

$$EP \cap JP \neq \Phi \tag{6.1}$$

where Φ represents the empty set, and $\cap$ indicates intersection. Goodman and Shi (1985) proved that the necessary and sufficient condition for a block to be finite is that

$$EP \cap JP = \Phi \tag{6.2}$$

Therefore all the united blocks of JP 11 are not finite. Unless new crack growth is considered, all such blocks are consequently safe.

(b) *JP = 01.* Any block bounded by joints and free surfaces such that it is simultaneously above joint plane 1 and below joint plane 2 has JP 01. Figure 6.2a shows a united block of JP 01. As in the previous case, EP is the half-plane under the horizontal line through the origin in Figure 6.2b. In this figure it can be seen that EP and JP 01 have common rays and therefore Equation 6.1 applies and all the united blocks of JP 01 are not finite. Such blocks are therefore not removable and cannot be key blocks; they are safe.

(c) *JP = 10.* Any united blocks with boundaries such that the block is in the lower half plane of joints of set 1 and in the upper half plane of joints of set 2 is a united block of JP 10. An example is shown in Figure 6.3a. Figure 6.3b shows that EP and JP 10 have common rays. Therefore all the united blocks of JP 10 are infinite and cannot be key blocks. The united block ruled in Figure 6.3a is infinite and intuitively one concludes that all related blocks are also infinite.

JP 00. United blocks of JP 00 are in the upper sides of joints of both sets 1 and 2. Two examples from the infinity of such blocks are shown by the ruled regions in Figure 6.4a and b. Figure 6.4c shows EP and JP 00, from which it can be seen that EP and JP have no common rays. Therefore Equation 6.2 applies and all of the united blocks of JP 00 are finite. Although this conclusion can be drawn intuitively in this simple, two-dimensional example, intuition is not so revealing in three-dimensional

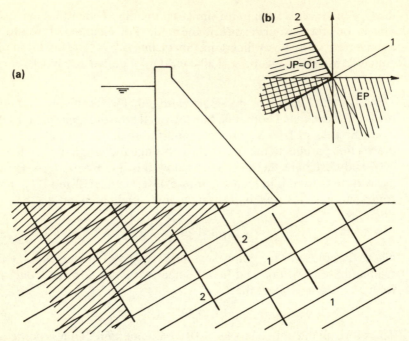

Figure 6.2 (a) The ruled region shows one of the infinite blocks of JP 01; (b) JP 01 has a non-empty intersection with EP.

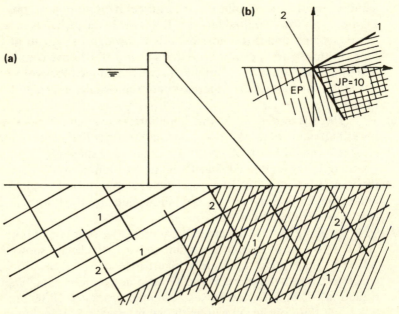

Figure 6.3 (a) The ruled region shows one of the infinite blocks of JP 10; (b) JP 10 has a non-empty intersection with EP.

cases. However, the application of Equations 6.1 and 6.2 in three dimensions are no more difficult, except that Figure 6.4c would be replaced by a stereographic projection of spherical polygons. Examples are given later (see Fig. 6.23a).

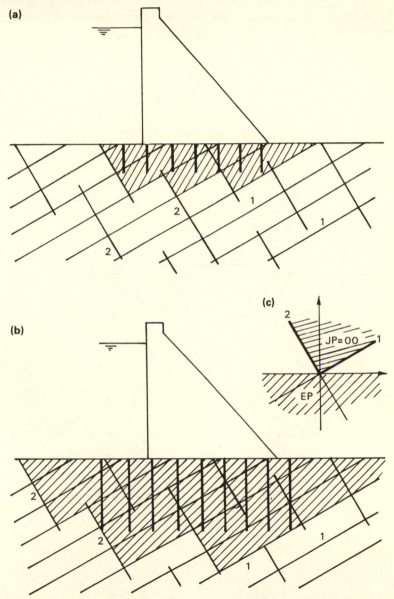

Figure 6.4 (a) The ruled region shows one of the finite blocks of JP 00 and a system of foundation anchors that makes this complex block integral; (b) the ruled region shows a larger finite block of JP 00, held by longer anchors; (c) JP 00 has no intersection with EP.

Figure 6.5a shows a united key block belonging to JP 00 slipping along joints of set 1. The direction of slip is a vector that can be drawn as a ray on the JP diagram, as shown in Figure 6.5b. A united key block with JP 00 can move along any direction belonging to JP 00 and JP may be thought of as the collection of all rays parallel to permissible movement directions. All united blocks of JP 00 are removable and, if the discontinuities are long, the number of such blocks is infinite. Such blocks can be very large. It is

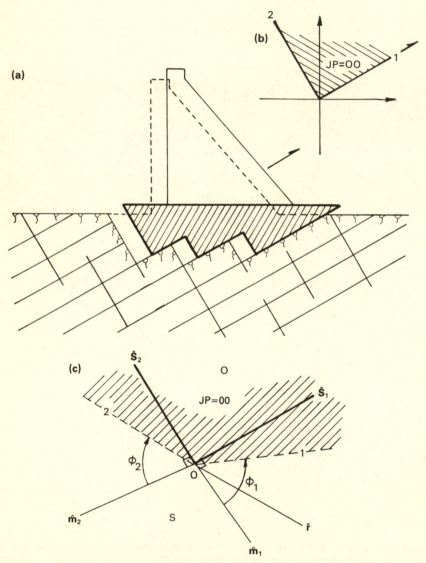

Figure 6.5   (a) A sliding, dilating mechanism for one block of JP 00; (b) The sliding direction is a direction of JP 00; (c) stability analysis for all sliding directions of JP 00.

nevertheless possible to discuss the stability of *all* such key blocks by introducing the concept of a **minimum stable key block**. First we examine the stability analysis itself.

6.5 Stability analysis of united key blocks

Consider again the united key block that slides as in Figure 6.5a. In Figure 6.5c, the angle between lines $\hat{S}_1$ and $\hat{S}_2$ is JP 00. Let $\hat{m}_1$ and $\hat{m}_2$ be vectors normal to $\hat{S}_1$ and $\hat{S}_2$ respectively and pointed away from the JP, as shown in Figure 6.5c. The shaded area of Figure 6.5c is enclosed within rays inclined ϕ_1 and ϕ_2 respectively with $\hat{m}_1$ and $\hat{m}_2$, as shown, the angles ϕ_1 and ϕ_2 being the angles of friction for sliding on planes 1 and 2. The modes of failure of the united key blocks of JP 00 depend on the direction $\hat{r}$ of the resultant force on the block.

(a) *If $\hat{r}$ is in region 0*, within JP 00, the block will open from each joint and 'pull-out' or 'lift' without shear along the boundary joints. (This is not strictly true if the joints are dilatant, in which case the region of lifting is reduced in angular extent by roughness angle *i* from each bounding ray of the JP.) The direction of movement of the united block will be parallel to $\hat{r}$.

(b) *If $\hat{r}$ is in region 1*, between $\hat{m}_1$ and $\hat{S}_1$, then the block will slide along the boundary joints of set 1 in direction $\hat{S}_1$. As a result of sliding, the bounding joints of set 2 will open. If the resultant force is directed outside the shaded area, there is sufficient frictional resistance to prevent slip and the block will be safe. If the resultant force direction, $\hat{r}$, is a ray inside the shaded area, available friction can not create equilibrium and the block will slide. The direction of the resultant force will then move to the position of the limiting value of ϕ_1.

If $\hat{r}$ is in region 2, united blocks having JP 00 will slide along the boundary joints of set 2 along direction $\hat{S}_2$. As a result of sliding, the boundary joints belonging to set 1 will open. In region 2, if the resultant force is directed outside the shaded area, the available friction is sufficient to prevent sliding; if the resultant force is directed along a ray inside the shaded area, the united block will slide and the direction of the resultant force will rotate to the ray, making a limiting angle ϕ_2 with m_2.

If $\hat{r}$ is directed in the region S between $\hat{m}_1$ and $\hat{m}_2$, there is no direction in which any block of JP 00 can slide. All united blocks of JP 00 will then be stable. By the **sliding force** we mean the vector sum of shearing forces and resisting forces in the direction of sliding, with shear forces reckoned positive in the direction of impending motion. The sliding force is negative for any united block having JP 00 when the direction ($\hat{r}$) of the resultant force is in the unshaded area of Figure 6.5c; similarly the sliding force is positive when the direction ($\hat{r}$) of the resultant force is in the shaded area of Figure 6.5c.

6.6 The minimum stable key block

There are an infinite number of united blocks of JP 00 and some of these are very large. It is impossible to support all of these blocks. In the case of the dam foundation of Figure 6.4, the resultant force ($R\hat{r}$) is the sum of the weights of the dam and the block, and forces caused by water pressure and inertia. If the united key block is very large, its weight will dominate all the other forces and $\hat{r}$ will be close to the direction of **W**. Since **W** is downward, $\hat{r}$ will then be in the unshaded area of Figure 6.5c, and such key blocks will be safe.

For a given set of friction angles, there will be a smallest united key block such that the sliding force is zero or negative. This **minimum stable key block** includes smaller blocks that are not safe without support. As shown in Figure 6.4, it is possible to consolidate a maximum unstable key block such that any remaining key block in the dam foundation will be larger and therefore have $\hat{r}$ directed in the safe region of Figure 6.5c. After the consolidation of the maximum unstable key block by anchoring, all the united key blocks of JP 00 will then be stable and because 00 is the only JP having removable blocks, all of the infinite united blocks of the dam foundation will be safe.

6.7 Clamped blocks

Blocks having parallel sides are termed **clamped blocks** because the presence of any roughness inhibits block slip. If such a block tends to move along a direction in the parallel faces, the incipient dilatancies of the opposed faces will be suppressed by the development of additional normal forces, which tend to clamp the block in place.

Figures 6.6 and 6.7 show clamped united blocks for the dam foundation previously considered. In Figure 6.6a, the block is always on the upper side of joints of set 2 and it is also bounded by parallel faces of joint set 1. Such a block is simultaneously in the upper and lower sides of the boundary joints of set 1; this condition will be represented by the digit 3 in the JP code. Accordingly, the united block shown in Figure 6.6a has JP 30. As indicated in Figure 6.6b, the graph of JP 30 is a ray directed entirely outside the EP for the dam foundation. Thus, Equation 6.2 applies and all of the blocks of JP 30 are finite. However, because they are clamped between the joints of set 1, they are safe unless these joints can be shown to have no roughness or to have thick clay filling, or the rock mass or abutments are so deformable that rock mass contraction under normal stress can accommodate the expectable dilatant tendencies.

Figure 6.7a shows a united block that is simultaneously in the upper and lower sides of the boundary joints belonging to both sets 1 and 2. The JP for such a block is 33. As shown in Figure 6.7b, the graph of this JP is the point

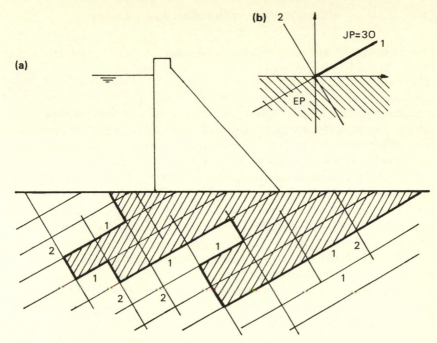

Figure 6.6 (a) The ruled region shows one of the finite blocks of JP 30; (b) JP 30 has no intersection with EP.

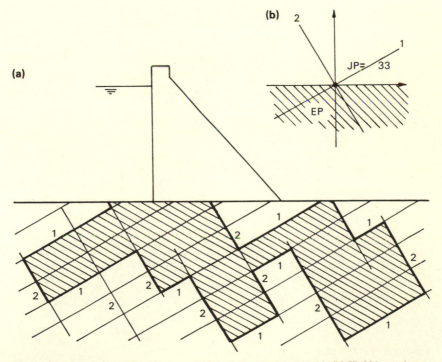

Figure 6.7 The ruled region shows one of the finite blocks of JP 33; (b) JP 33 is a point.

of intersection of rays parallel to set 1 and to set 2, which is exactly the origin (0,0). Therefore, for this class of blocks,

$$JP = \Phi \qquad (6.3)$$

meaning that the blocks belonging to this JP are finite without their intersection by a free surface. Any such block is tapered, i.e. there is no direction of block motion that does not cause collision with an abutment. All these blocks are therefore not removable; hence they are safe.

Figure 6.8 shows the rays corresponding to JP 03, JP 31, and JP 13. All of the united blocks belonging to these JPs are clamped and therefore safe, except for the special cases of smooth or filled joints or highly deformable rock mass, as noted previously. JPs 31 and 13 yield only infinite blocks, whereas JP 03 yields finite blocks.

From these examples for the dam foundation shown in Figure 6.4, the importance and utility of the joint pyramid is established. The use of the JP is simple, yet it provides the essential information for geometric and stability analysis. It is sufficient and far more convenient to examine the JP than to try to draw the individual complex united blocks.

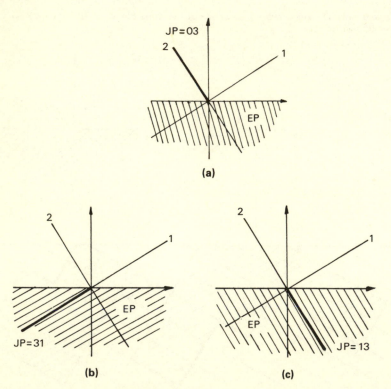

Figure 6.8   (a) JP 03 has no intersection with EP; (b) JP 31 has a non-empty intersection with EP; (c) JP 13 has a non-empty intersection with EP.

6.8 Definitions of united and convex blocks

A block is a connected area (or volume in the three-dimensional case) bounded by joints and/or free planes in any combination. In Figure 6.9a, the

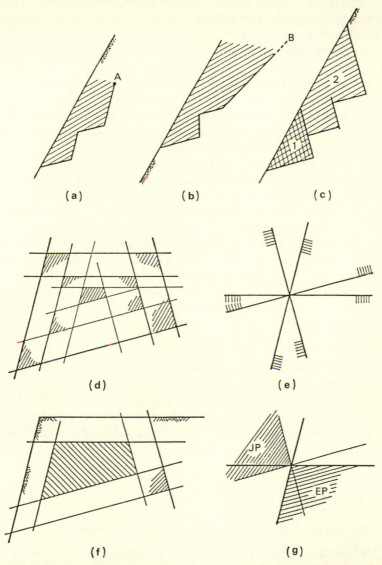

Figure 6.9 (a) The shaded area is not a block, according to the definition; (b) the shaded area is an infinite block; (c) the shaded area is a finite block; (d) a convex block reduces to a point when all its sides are moved inward; (e) the intersection of all moved half-spaces of (d) is a point; (f) moving all the joint half-spaces of a convex block towards the centre of the block yields the shaded region termed JP, whereas moving all the free plane half-spaces of a convex block towards the centre of the block yields the shaded region called EP; (g) JP and EP have no intersection.

shaded area is connected with the remainder of the plane beyond the termination of the joint at point A. Therefore the shaded area of Figure 6.9a is not a block.

An **infinite block** is shown in Figure 6.9b. That the shaded area is infinite in this case is established by the divergence of its boundaries. That the shaded area is a block is established by the absence of any connection between the shaded area and the rest of the plane.

A **united, finite block** is shown in Figure 6.9c by the union of the shaded areas 1 and 2.

A **convex, finite block** is indicated by the shaded area 1 on Figure 6.9c. A convex region is one in which the shortest route between any two of its points lies entirely inside the region. Convex blocks are easier to work with than non-convex blocks. Therefore united blocks will be decomposed into convex blocks for the purpose of stability analysis.

Figure 6.9d shows a simple, four-sided convex block, established by the area common to four half-planes limited by four lines. As we move all the lines towards the centre of the block, the block reduces to a point $(0,0)$ (Figure 6.9e). It is also true that the common area of the four half-planes shrinks to the same point. Two of these four half-planes are limited by free surfaces and two are limited by joints, as shown in Figure 6.9f. If we move the four bounding lines to $(0,0)$ there remains an area common to the two half-planes limited by the free surfaces; this common area is the excavation pyramid, EP. Similarly, the area common to the half-planes limited by the joints is the joint pyramid (JP). The area common to JP and EP is only $(0,0)$, which includes no rays, i.e.

$$\text{EP} \cap \text{JP} = \Phi \tag{6.2}$$

Equation (6.2) is the criterion of finiteness for a convex block.

A united block can be decomposed into convex sub-blocks whose union equals the original united block. The boundary of each convex block will be an extension of a boundary segment of the united block such that both the convex block and the united block occupy the same side of the boundary segment. Examples are shown in Figure 6.10.

6.9   Types of united blocks

Four different types of united blocks are depicted in Figure 6.10.

(1) **Joint blocks**, as shown in Figure 6.10a, are regions enclosed entirely within bounding joint planes. The example of Figure 6.10a can be subdivided into three convex blocks. In every one of these, the JP is empty. In general, for a union of n convex joint blocks,

$$\text{JP}_i = \Phi, \qquad i = 1, 2, \ldots \tag{6.4}$$

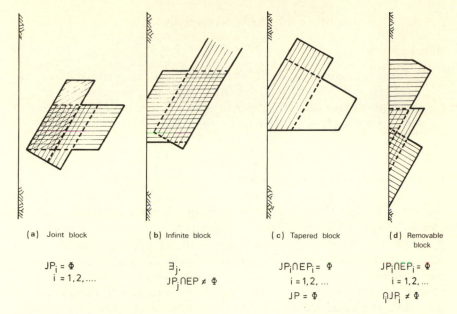

(a) Joint block   (b) Infinite block   (c) Tapered block   (d) Removable
                                                                                          block

$$JP_i = \Phi$$
$$i = 1, 2, ...$$

$$\exists_j,$$
$$JP_j \cap EP \neq \Phi$$

$$JP_i \cap EP_i = \Phi$$
$$i = 1, 2, ...$$
$$JP = \Phi$$

$$JP_i \cap EP_i = \Phi$$
$$i = 1, 2, ...$$
$$\underset{i}{\cap} JP_i \neq \Phi$$

Figure 6.10   (a) A complex joint block; (b) a complex infinite block; (c) a complex tapered block; (d) a complex removable block.

(2) **Infinite blocks** have at least one infinite component block. An example is shown in Figure 6.10b, where an infinite block is formed as the union of a single finite block and a single infinite block. In any infinite block, there is at least one value of j such that JP_j has a non-empty intersection with EP_j, i.e.*

$$\exists_j \quad \text{such that} \quad JP_j \cap EP_j \neq \Phi$$

(3) **Tapered blocks** have no direction of motion that does not tend to cause closing of a boundary joint. In Figure 6.10c, the united block is divided into two convex sub-blocks. The criterion for tapering is that the joint pyramid of the united block be empty. If each convex sub-block is finite,

$$JP_i \cap EP_i = \Phi, \quad i = 1, 2, \ldots$$

The criterion for tapering is

$$JP = \underset{i}{\cap} JP_i = \Phi \tag{6.5}$$

(4) **Removable blocks** are represented by the example in Figure 6.10d, showing a united block composed of the union of three convex sub-blocks. Since each convex sub-block is finite,

$$JP_i \cap EP_i = \Phi, \quad i = 1, 2, \ldots$$

* $\exists_j$ means 'there exists a value of j'.

and since none of the sub-blocks is tapered,

$$JP = \cap \, JP_i \neq \Phi \qquad (6.6)$$

JP is the set common to all JP_i. The united block can be moved along any direction of JP, so if $JP \neq \Phi$ the united block is removable.

Although four different types of united blocks have been discussed above, only the removable blocks require stability analysis.

6.10   Utility of the joint pyramid

As noted, a single JP can represent an infinite number of complex, united blocks. The properties of a JP can be computed easily as every JP is a convex pyramid. Knowledge of the properties of the JP provides decisive information about the stability of all the united blocks. For any given excavation, the JP allows one to determine the finiteness and removability of all of its blocks. For a given direction of the resultant force, the JP permits one to determine the modes of failure and the directions of sliding for all of its blocks. Several special cases will be examined.

6.11   Stability conditions when EP is of maximum size

Figure 6.11a shows a finite united block intersecting the roof and two opposite walls of a tunnel. This block can be subdivided into three convex sub-blocks by continuing the lines of the walls and roof as shown. Since each of the three convex blocks is finite, then for each i

$$EP_i \cap JP_i = \Phi, \qquad i = 1, 2, 3 \qquad (6.7)$$

and

$$EP_i \cap JP = \Phi, \qquad i = 1, 2, 3 \qquad (6.8)$$

Since the entire united block is finite,

$$(EP_1 \cup EP_2 \cup EP_3) \cap JP = \Phi \qquad (6.9)$$

EP_1, EP_2, and EP_3 are shown in Figures 6.11b, c, and d, respectively. The union of these is the whole plane, as shown in Figure 6.11e. Accordingly, Equation 6.9 can be satisfied only if $JP = \Phi$. Therefore, the united block shown must be tapered. In general, any blocks that occupy opposed faces of a tunnel or underground chamber must be tapered. All such blocks are safe.

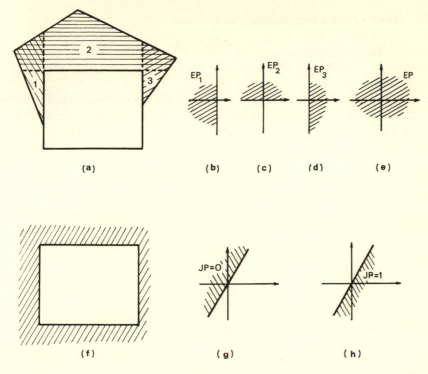

Figure 6.11 A tapered block in a tunnel: (a) the tapered block is seen as the union of convex sub-blocks 1, 2, and 3; (b), (c), and (d) the corresponding EP of each sub-block; (e) EP of the whole block is the union of the EP_i; (f) a tunnel with one joint set where the joints are very dense and very long; (g) and (h) are the JPs of this joint set 0 and 1 respectively.

6.12   Stability conditions when JP is of maximum size

The largest possible joint pyramid arises in the case of only one set of joints. A tunnel through a rock mass with one joint set is depicted in Figure 6.11f. There are only two possible JPs in this case – JP = 0 and JP = 1. Both of these correspond to half-planes, as shown in Figures 6.11g and h. Any EP_i in the tunnel or underground chamber is equal to a half-plane as well. Since any two half-planes through a common origin include an angle, for any convex sub-block

$$EP_i \cap JP_i \neq \Phi$$

Accordingly, there are no finite blocks. Without rock material failure there can be no failures in the tunnel, even if the joints are very closely spaced and of infinite extent.

6.13 Blocks with a compound free surface

A united block can derive its non-convexity from the joint system, from the free surfaces, or from both. Several examples with complex free surfaces will now be considered.

Figure 6.12 shows a convex slope in which the rock mass occupies the convex area limited by lines F_1 and F_2. By extending the joints, the united block is decomposed into two finite, convex sub-blocks; therefore,

$$JP_i \cap EP_i = \Phi, \qquad i = 1, 2 \tag{6.10}$$

EP_1 is the same as EP_2 in this case, but JP_1 differs from JP_2, as shown in Figures 6.12b and c. As shown in Figure 6.12d,

$$JP \neq \Phi \quad \text{and} \quad JP \cap EP = \Phi \tag{6.11}$$

where

$$EP = EP_1 \cup EP_2 \tag{6.12}$$

and

$$JP = JP_1 \cap JP_2 \tag{6.13}$$

Thus, there are key blocks of JP in the convex slope limited by F_1 and F_2.

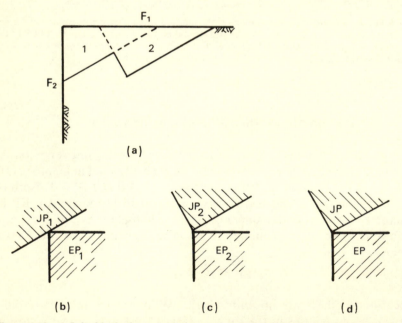

Figure 6.12 A finite block of an outside (convex) corner: (a) the block is decomposed into two convex sub-blocks; (b) EP and JP for sub-block 1; (c) EP and JP for sub-block 2; (d) EP and JP for the complex block where EP is the union of EP_i and JP is the intersection of JP_i.

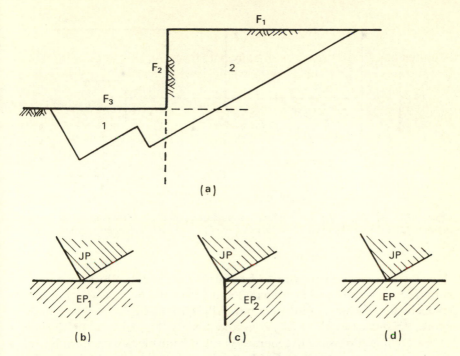

Figure 6.13 (a) Decomposition of the block into the two sub-blocks shown is not satisfactory because the sub-blocks are not themselves convex. But the free surface can be decomposed, as shown, into a half-space, F_3, and a convex corner, F_2 and F_1. (b) and (c) show JP and EP for these two parts of the free surface; (d) shows JP and EP for the whole complex block.

Figure 6.13 shows a slightly more complicated case, with a non-convex slope limited by lines F_1, F_2, and F_3. The decomposition of the united block is accomplished by extending free surface lines F_2 and F_3 as shown. JP_i is the same for both component blocks, whereas EP_1 and EP_2 are now quite different; EP_1 is a half-plane and EP_2 is the right angle common to two half-planes, as shown in Figures 6.13b and c respectively. For this case, Figure 6.13d determines that

$$JP \cap EP_i = \Phi, \qquad i = 1, 2 \tag{6.14}$$

and

$$JP \neq \Phi \tag{6.15}$$

The conclusion is that the free boundaries F_1, F_2, F_3 admit key blocks.

Now consider the more complicated case of Figure 6.14a, with a free surface created by four lines, F_1, F_2, F_3, and F_4, as shown. The latter extends infinitely so that there is no upper surface corresponding to F_1 on the left side of the excavation. Assume that there is a united block of the JP that has these four free surfaces as boundaries. This united block can be decomposed into

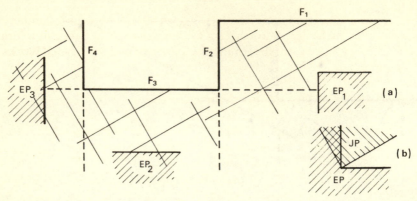

Figure 6.14 (a) The complex free surface can be decomposed into three convex free surface components; (b) EP is the union of the three EP_i components; it has an intersection with JP.

three united blocks having convex free surfaces depicted. Block 1 is formed with free surfaces F_1 and F_2; Blocks 2 and 3 have F_3 and F_4 as free surfaces, respectively. The excavation pyramids EP_1, EP_2, and EP_3 are shown in the figure. In this case, $JP \cap EP_3 \neq \Phi$ and since $EP \cup_i EP_i$, $JP \cap EP \neq \Phi$. Therefore it is impossible to generate a united finite block with the indicated JP and the free surfaces F_1 to F_4. If a horizontal top surface, F_5, is added to the previous case, as shown in Figure 6.15, a finite united block can exist. The block shown in Figure 6.15a is decomposed by extending the free surfaces F_2, F_3, and F_4 as shown. There are now three EP_i components and the union of these produces a half-plane which has no intersection with JP.

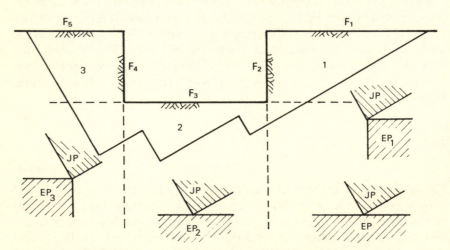

Figure 6.15 As in Figure 6.14, the complex free surface is decomposed into three convex parts; EP for the complex block is the union of the EP_i; it has no intersection with JP.

In all the examples of united blocks that we have discussed, it is unnecessary to study the complex, united blocks themselves. It is sufficient to draw the JP$_i$ and EP$_i$ in order to determine the existence of finite, united blocks.

6.14   Support of the maximum unstable block

The concept of a minimum stable block and a maximum unstable block was introduced previously. This concept makes it possible to design supports for definite and complete stability analysis of an infinite block system. In choosing supports, one can adopt either of two philosophies. In the first, the support is assumed to apply a certain force, which when added to the other forces on a block turns their resultant into the safe zone. In the second approach, one assumes that the support acts as a constraint on the joints that it crosses; a joint that crosses the support cannot slip and, therefore, cannot bound a key block. To achieve the intention of the first philosophy, the supports have to be installed with prestress or be apportioned to develop sufficient passive force following a *permissible* displacement. Bolts or cables are typical instruments of this approach. To achieve the intent of the second design approach, the support has to completely constrain any crossing joint from shearing. A concrete pier or keyway, or a concrete or steel lining grouted against the rock surface, can function in this way.

Figure 6.16a shows a concrete pier in a rock foundation. If the pier does indeed constrain all crossing joints, no block whose boundary crosses it can be a key block. However, any united key block that contains this pier is unaffected by it. Such a block must contain a minimum key block; if this

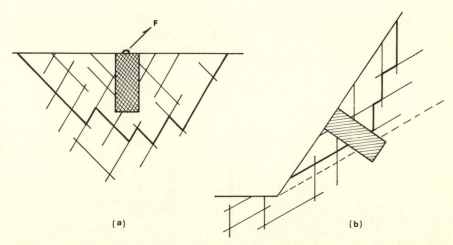

Figure 6.16   The concept of a support as a kinematic barrier, preventing movement of small blocks; (a) application in a foundation; (b) application in a surface cut.

minimum key block is stable under the resultant force, all of the larger blocks must also be stable. In the figure, the resultant is the vector sum of the weight of the block and of the pier and an upward structural pull **F**.

Figure 6.16b shows a concrete pier installed in a rock slope. In this excavation, there is a maximum key block. The pier is so long that the minimum block that contains it is larger than the maximum key block of the excavation lacking a pier. Therefore, the entire slope must be stable. The pier can be likened to the keystone of a big three-dimensional puzzle.

6.15   The existence of a maximum key block and its use

For any given JP, the union of two of its united key blocks is a larger united key block. In many excavation configurations, there is a maximum key block that contains all the united key blocks of a given JP. The existence of a maximum key block is useful in the design of rock supports, as illustrated by the previous example.

The maximum key block concept can be applied, despite our general inability to see all its bounds in any excavation. Figure 6.17a illustrates this point. The free surface following the excavation of the rock slope has an

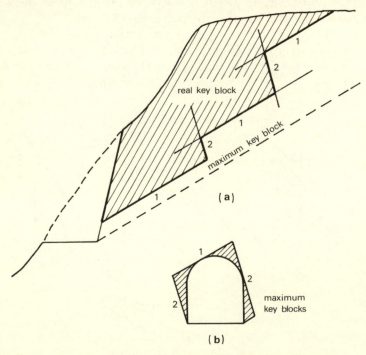

Figure 6.17   Maximum key blocks for two roadways of identical width in the same rock mass: (a) surface excavation; (b) underground excavation.

associated maximum key block of enormous dimensions, whose movement could cause an engineering catastrophe. The orientation and configuration of this maximum key block are known prior to excavation, even if particular component key blocks are not visible from the initial surface. It is too late to address the hazards of the excavation after the cut has been bottomed out, but the engineer can use the maximum key block concept to organize a safe sequence of cutting and support. Or a tunnel can be considered to replace the surface excavation. The maximum key block of a tunnel is likely to be far smaller than that of a surface cut through the same geology. The rock cut in Figure 6.17a has the very large key blocks shown, while the tunnel through the same rock mass, with identical roadway widths, has only the two smaller key blocks shown in Figure 6.17b. It will be easier to support the tunnel to a given probability of safety.

6.16   The maximum key blocks of tunnels

Tunnels have complex free surfaces as they are formed by unions of curved cylinders arranged about a generally directed and inclined axis. The characteristic cross section of a tunnel is a loop of almost any shape in the plane normal to this axis. Nevertheless, the maximum key blocks of the most general tunnel can be determined by the following simple procedure.

(a) Determine the direction and inclination of the tunnel's axis and the dip and dip direction of the plane perpendicular to it.
(b) Compute the three-dimensional joint pyramid JP and project its edges into the plane perpendicular to the tunnel axis (Fig. 6.16a).
(c) Move the vertex of the JP in the tunnel section until the extreme lines of its projection become tangent to the tunnel periphery, as in Figure 6.18b. The region thus enclosed is the maximum key block of the JP.

The intersection of the translated JP with the tunnel is actually a three-dimensional volume of which we have produced only a projection in the tunnel section. The actual maximum key block for the case studied in Figure 6.18a is shown in Figure 6.18e.

The maximum key block region constructed in Figure 6.18b in the tunnel section is the upper bound for all united key blocks that may occur from this JP in the tunnel. The whole region is occupied if all possible key block positions are realized, in the case of very long and very closely spaced joints. Considering the finite length and spacing of real joint systems, the maximum key block region can be reduced, as shown in Figure 6.18c. Taking advantage of the information about lengths and spacings of joints, support of only the shaded area of Figure 6.18c will then stabilize all the key blocks of the JP.

The maximum key block region can be reduced by considering compressive stresses to act across it. These can originate from the state of stress about

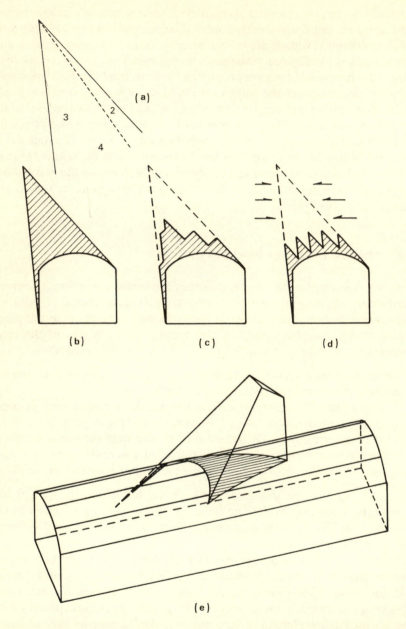

Figure 6.18 (a) Projection of a three-dimensional joint pyramid into the plane perpendicular to a tunnel; (b) maximum key block envelope; (c) key block envelope with finite joint length and spacing; (d) key block envelope with finite joint length and spacing and tangential stress around the tunnel; (e) three-dimensional view of a curved key block from the same JP.

the tunnel or from incipient block movement if the bounding joint planes are rough, providing that the vertex of the maximum key block region is sufficiently narrow. When the JP has almost parallel sides, the maximum key block tends to be very large; however, such a block has a very narrow angle and any roughness will promote an increment in normal stress as a result of a movement tendency for the block. Accordingly, the maximum key block region will be greatly reduced, as illustrated by Figure 6.18d.

6.17 Choice of the direction for a tunnel

Quite often the joints of one set will be very long, and closely spaced; the obvious example is jointing parallel to bedding in rhythmically bedded sandstones and shales. If the joint planes are smooth and the compressive stress normal to them is small, it may be necessary to consider the potential movement of clamped, united blocks.

As an example, consider a dominant, smooth set of joints parallel to the planes of bedding, represented by the inclined great circle in the stereographic projection of Figure 6.19a. Since the JPs for clamped blocks parallel to this plane occupy both the upper and lower hemispheres of this great circle, all such JPs are segments of this circle. Three cases are shown in Figure 6.19.

(1) When the axis of the tunnel is parallel to line 1, close to the strike of the bedding, the maximum key block region is very large, as shown in Figure 6.19b. For this case, Figure 6.19c shows one large key block with parallel faces along the bedding.
(2) When the axis of the tunnel is parallel to the direction of line 2 in Figure 6.19a, the maximum key block region is considerably reduced, as shown in Figure 6.19d. Figure 6.19e shows a maximum clamped key block for a tunnel in direction 2.
(3) When the tunnel is directed parallel to a line through 3, in the direction of dip of the bedding, the maximum key block is even smaller, as shown in Figure 6.19f; Figure 6.19g depicts a maximum key block for the tunnel in direction 3.

In this example, if the joints parallel to the bedding are very smooth and the compressive stress is small, the great size of the maximum key block region for direction 1 suggests that it may be very difficult, and expensive, to insist on driving a tunnel in that direction. The potential savings from redirecting the tunnel to direction 3 are so great as to warrant an extra effort in this regard. We are not impressed by arguments that designers lack such freedom. The owning agency will be swayed by the cost penalties attached to incorrect orientations for underground works, for the extreme range in possible support and excavation costs accomplished by extreme swings in orientation is far greater than ranges of other cost items.

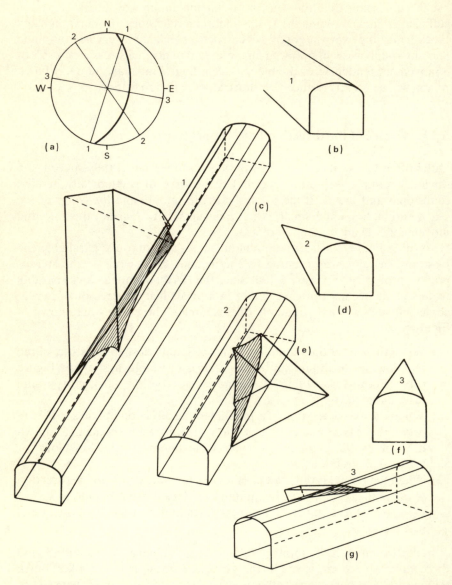

Figure 6.19 (a) Directions 1, 2, and 3 for a tunnel through a rock mass with an important repeated joint set, e.g. bedding joints, indicated by the great circle; (b) and (c) maximum key blocks envelope and a three-dimensional view of a curved block for a tunnel parallel to direction 1; (d) and (e) the same for a tunnel in direction 2; (f) and (g) the same for a tunnel in direction 3.

6.18   Complete key block analysis of tunnels

The following example considers an inclined tunnel parallel to direction $\hat{t}$ in the stereographic projection of Figure 6.20a. The four sets of joints in the rock mass are represented in the stereographic projection by the four great circles. Blocks may be formed by the intersection of either the upper or lower half-space of each of the four joints, so the number of possible joint pyramids is $2^4 = 16$. For each of these joint pyramids, we shall draw the maximum key block region and discuss the stability of its united key blocks. We assume that the resultant acting on each key block is the weight of the key block alone; all the joints are assumed to be very long and closely spaced.

In Figure 6.20b, the maximum key block regions for each JP are drawn for a given tunnel shape, and all the drawings are superimposed on the stereographic projection of Figure 6.20a. Each of these inset figures is discussed in Table 6.1. The results discussed in Table 6.1 indicate that although there are 16 joint pyramids associated with the four joint sets only three key block regions are important, even though we assumed the joints to be long and closely spaced. These are formed by JPs 1001, 1011, and 1101. All the dangerous key block regions are on the left side of the tunnel, so support on the right side of the tunnel is less important, if necessary at all. If the initial stress in the rock mass is characterized by horizontal stress about equal to the vertical stress ($K \simeq 1$), the periphery of the excavation will be in a state of compression and the large united key blocks will be stabilized.

In general, all joints are not equally well developed and not all JPs equally significant. From a design point of view, it is important to seek solutions that minimize the chances of encountering key blocks belonging to the most important JPs. A formal analysis of this problem for any direction of resultant, termed **mode analysis**, is presented in Goodman and Shi (1985). In problems where gravity supplies the main contribution to the active resultant forces on blocks, the most troublesome blocks lie above the most critical joints where they are in a position to slide on them. Therefore an approximation to the list of most critical JPs is the list of JPs having the symbol '0' in the positions of the most significant joint sets, i.e. the joints that are longest, and smoothest. To find a direction for a tunnel that assigns only safe maximum key block regions to most of these JPs, choose a direction for $\hat{t}$ that lies near the intersection of dangerous JPs. Placing $\hat{t}$ near the intersection of two joints cancels the four JPs that share the intersection as a common corner. This is also true of $-\hat{t}$, which cancels the four cousins of these JPs. (A cousin of a JP is a new JP formed by interchanging every 1 and 0 of its code; for example, 1001 is the cousin of 0110.)

In considering the results of a complete key block analysis of a tunnel, the portion of the tunnel periphery that merits especially careful support is the zone that is common to the maximum key block regions. If it is possible to determine the state of stress around the tunnel, it is wise to treat specially any zone that combines dangerous maximum key block regions and low tangential stress.

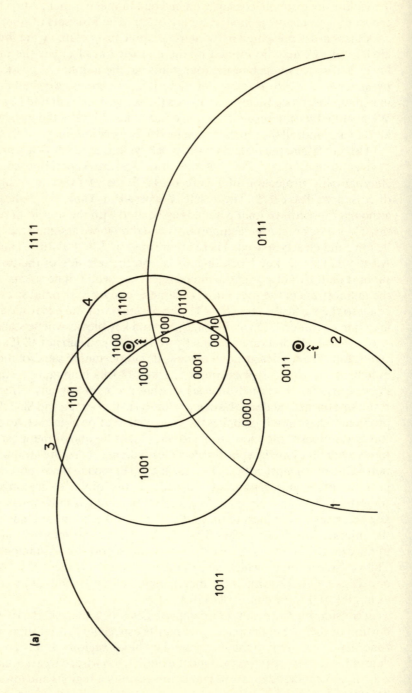

(a)

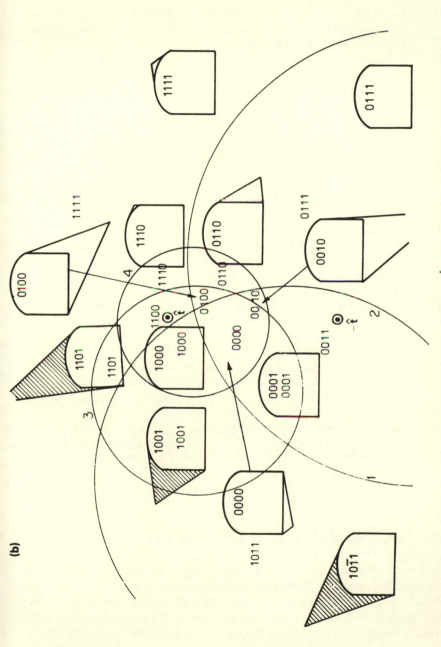

Figure 6.20 (a) Stereographic projection of joint pyramids and the axis ($\hat{\mathbf{t}}$) of a tunnel; (b) maximum key block envelope of every JP of Figure 6.20 (0011 and 1100 have no key block envelope).

Table 6.1   Maximum key block regions for rock mass of Figure 6.20a.

JP	Discussion of removability and stability
0000	removable, but there is no mode of sliding; all blocks are safe
0001	no maximum key block region because of the tunnel shape; so there are no removable key blocks of this JP
0010	removable, but there is no mode of sliding; all blocks are safe
0011	since the opposite $(-\hat{t})$ to the tunnel axis is in this JP, all united blocks of this JP are infinite and therefore safe; however, there may be blocks of 0011 in the face of the tunnel
0100	removable, but there is no mode of sliding; all blocks are safe
0101	this JP does not appear in the stereographic projection; it is empty and therefore all blocks with this JP are tapered and safe
0110	removable, but the lower limit line of the maximum key block region is almost horizontal, meaning that the sliding direction has a very small dip angle (*the sliding direction cannot be steeper than this line*); a very small friction angle is therefore sufficient to stabilize any united key block
0111	removable, but there is no mode of sliding; all blocks are safe
1000	there is no maximum key block region because of the shape of the tunnel, so no removable blocks will exist for this JP
1001	removable, and the lower limit line of the maximum key block region dips steeply; finite, united key blocks of this JP will probably require support
1010	this JP does not appear in the stereographic projection; it is therefore empty and any key blocks formed from it are tapered and accordingly safe
1011	removable, with a large maximum key block region whose lower limit dips steeply; key blocks of this JP will probably need support
1100	this JP includes the projection $\hat{t}$ of the axis of the tunnel; therefore all the united blocks of this JP are infinite and consequently safe
1101	removable, with a large, maximum key block dipping steeply; key blocks of this JP will need support
1110	removable, with a very small maximum key block because of the shape of the tunnel; all of the key blocks of this JP will be very small and may be unimportant – they can be stabilized with very little support
1111	removable, but with a small key block region in the roof; a small amount of support will stabilize key blocks of this JP

6.19   United key blocks between parallel tunnels

In a complex engineering system necessitating parallel tunnels or shafts, the designer must choose optimum separation distance to effect an economical and safe solution. In jointed rock, this decision can be supported by key block analysis because the potential movement of key blocks in the pillar between tunnels can undermine the excavation system. An example (Fig. 6.21) illustrates how the key block analysis applies to this problem; the

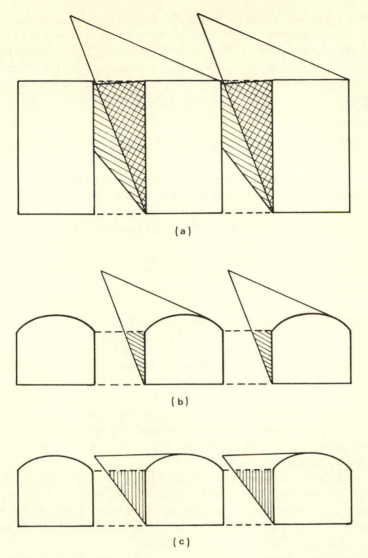

Figure 6.21 (a) Closely spaced, high tunnels in the rock mass illustrated in Figure 6.20 have intersecting maximum key block envelopes allowing the possibility of pillar collapse; (b) and (c) satisfactory spacing of parallel tunnels to avoid any overlap of maximum key block regions.

example is based upon the same tunnel direction and four sets of joints previously studied.

Of the 16 possible JPs, Figure 6.20b shows that only 1001 and 1011 have large maximum key block regions in the rock mass between parallel tunnels. These have been transferred to Figure 6.21. If the tunnels are as high as shown in Figure 6.21a, and spaced as indicated, one would need to support the entire region that is shaded on this figure. It is possible to choose tunnel shapes and separations to make it impossible for the maximum key block regions to undermine the whole of the intertunnel pillar, as shown in Figures 6.21b and c for each of the critical JPs. The global stability analysis for selecting tunnel spacing in a rock mass with four sets of joints is thus simply analysed.

6.20   The complete stability analysis of an underground chamber

In the sense that an underground chamber has planar walls, it is simpler than a tunnel. But even the simplest boxlike gallery has at least four walls, a roof, and a floor, i.e. six free planes, as well as numerous edges and corners. We discuss here the complete analysis of such a chamber considering all united key blocks.

It was shown previously that removable united blocks cannot occupy two opposed faces of an underground opening. The combinations of free planes that can admit a single, united key block are as follows.

(a) A united key block cannot have more than three free planes of a prismatic underground chamber. The periphery is comprised of three pairs of free planes; hence, if there were more than three free planes in one united key block, that block would include one pair of opposite free planes, but this cannot be.

(b) From each pair of parallel free planes, we can choose only one, and since there are only three pairs of free planes, there are as many as $2^3 = 8$ combinations. These eight combinations of three free planes are the eight corners of the chamber and the corresponding key blocks are called **key blocks of the corners**.

(c) The total number of free plane combinations, choosing two at a time, is $C_6^2 = 15$. However, three combinations of these planes involve opposed pairs; a key block cannot occupy opposed faces of the underground chamber. Therefore there are only 12 combinations of free faces that can admit united key blocks. These are the 12 edges of the chamber and the key blocks that contain them are called **key blocks of the edges**.

(d) There are six free surfaces in the underground chamber – four walls, a roof, and a floor. Each of these can have united key blocks.

The number of possible free plane combinations that form united key blocks is therefore quite large: $8 + 12 + 6 = 26$. All of these possibilities can be checked, because the existence of a united key block in any of them is governed by the same set of conditions:

$$JP \cap EP = \Phi \quad \text{and} \quad JP \neq \Phi$$

We shall give the EP for all 26 combinations of free planes. With this information, for any particular system of joints the existence of all key blocks can be found directly from the stereographic projection.

Figures 6.22a, b, and c present horizontal sections of the roof, walls, and floor of a prismatic underground chamber, as seen from above. In the stereographic projection, a vertical plane projects as a straight line. Taking a lower focal point (upper hemisphere) projection, a horizontal plane projects as the equatorial circle and the half-space above this plane is the area within this circle.

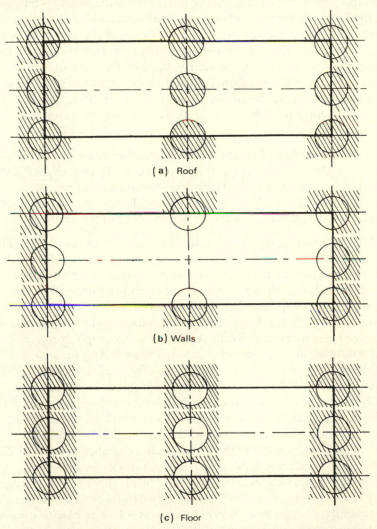

(a) Roof

(b) Walls

(c) Floor

Figure 6.22 (a) Stereographic projections showing EP for the four roof/wall edges, the four roof/wall/wall corners, and the roof alone; (b) stereographic projections of the EP for four walls and four wall/wall edges; (c) stereographic projections of EP for four wall/floor edges, four wall/wall/floor corners, and the floor itself for a prismatic underground excavation.

The EP for every one of the 26 combinations of free planes is shown by the shaded area of the stereographic projections overlaid on Figures 6.22a, b, and c.

Roof sections. The excavation pyramid for the roof is the half-space above the roof plane and is therefore represented by the shaded area inside the equatorial circle (Fig. 6.22a). The EPs for all of the roof edges are formed by the union of this half-space and the half-space of one wall. Thus, for example, the roof/wall edge on the north side of the gallery is represented by the EP that is the union of the upper half-space and the half-space north of the wall. It is projected, therefore, by the shaded area at the top centre of Figure 6.22a – an area uniting the region above the horizontal line and inside the equatorial circle. Similarly, the EP of a roof corner is the union of the area inside the equatorial circle with the outside angle of the corner. The similarity of the stereographic projections to the actual plan of each particular portion of the chamber makes it easy to remember the diagrams.

Wall and floor sections. Figure 6.22b presents similar stereographic projections for the excavation pyramids of the walls and edges (wall/wall intersections). Figure 6.22c shows the stereographic projections for the EPs of the wall/floor edges and the wall/wall/floor corners. The diagram in the centre of Figure 6.22c is the EP for the floor of the chamber.

All of these stereographic projections are used in the same way. Having determined the orientations of the different sets of joints, the JPs are defined as circular polygons on the stereographic projection. A removable finite key block is formed only when a JP has no intersection with the EP. Thus, if a JP lies entirely outside a shaded area in one of the stereographic projections of Figure 6.22, there are finite, removable blocks belonging to the edge, corner, or face represented by the particular stereographic projection.

For example, the four sets of joints of Figure 6.20 generate the JP 1101 which lies outside of the EP of a roof corner (Fig. 6.23a). Figure 6.23b shows the key block, which has the roof and two adjacent walls as free planes. Since the EP of the edge formed by these two walls is a component of the EP shown in Figure 6.23a, 1101 is also a potential key block of the wall/wall edge. United key blocks of the roof/wall/wall corner and of the wall/wall edge are both shown in relationship to the underground chamber in Figure 6.23c.

There has been considerable attention, in rock mechanics practice, to locating blocks and wedges in underground faces. However, the recognition of key blocks moving simultaneously into two or three adjacent faces is not as frequently discussed. It is possible to avoid such blocks with minor adjustments of an underground opening's orientation because the EPs of edges and corners are quite large. On the other hand, any block that satisfies the geometric requirements to be a key block of an edge or corner has a

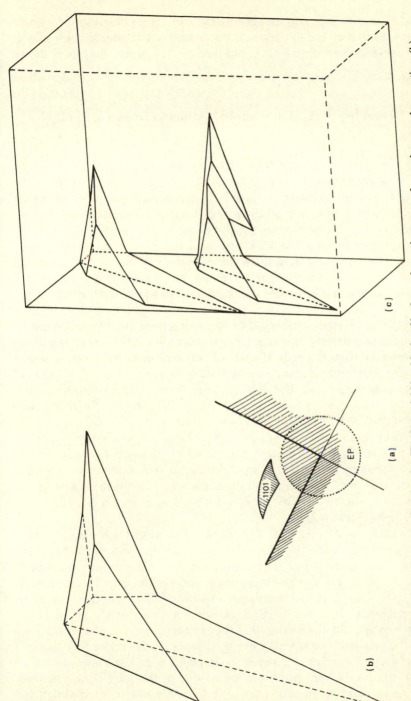

Figure 6.23 (a) The stereographic projection showing JP1101 to yield key blocks of both an edge and a corner of an underground excavation; (b) a three-dimensional view of a complex key block of JP1101 in the corner of an underground excavation; (c) three-dimensional key blocks of JP1101 of both the corner and the edge of the underground excavation.

(a)

(b)

(c)

maximum key block region that is considerably larger than that of a single free face. Such a united key block is potentially very dangerous therefore, and its avoidance needs to be assured through key block analysis as shown here.

6.21   Delimiting the real maximum key blocks from a map of joint traces

During the period of excavation, making use of photogrammetry, an engineering geologist can map the joints of an excavated rock surface. The traces drawn on such a map provide vital information about the nature of the excavation and a basis for prediction of what is to come in subsequent excavations. The intersections of the traces outline the faces of blocks, some of which are potential key blocks. Using block theory, the real maximum key blocks can be identified from the joint trace map by the following manual method. This method is like game-playing, but the stakes are real as the failure to locate a maximum key block can lead to a failure to protect lives.

(1) *Eliminate all 'trees'*. Figure 6.24a is a typical joint map. If a joint has no intersection or only one intersection point with other joints, it may be eliminated from the map. If a joint has two or more intersection points where other traces cross it, we shall have to remove the two 'dead-end' segments at each end. The dead ends are the line intervals from the ends of the joint trace to the nearest intersection point. Erase the two dead-end segments at each end of each joint trace.

  After eliminating the segments above there remains a slightly simpler map. The removal of some segments affects those that remain, so it is now necessary to repeat the steps above and to re-examine the simplified trace map, etc., iterating until there are no more changes to be made. Each line segment of this modified trace map is now a boundary of a polygon (Figure 6.24b).

(2) *Introduce the JP codes for key blocks*. The joint trace map is a two-dimensional section through a three-dimensional rock mass. From the key block analysis one determines the JPs that produce potentially unsafe, finite, removable blocks in the plane of the map. Recall that the JP codes 0 and 1 identify the upper and lower half-spaces of a joint plane respectively. In a non-vertical two-dimensional joint map, it is not obvious and not always true that the region above (i.e. towards the top of the map) of a joint trace belongs, in fact, to the upper half-space of the particular joint. However, it is possible to determine to which half-space of a joint plane the upper side of the joint's trace actually belongs. A simple method for doing this is presented in Goodman and Shi (1985). Having done this for each joint, we transfer the JP code into

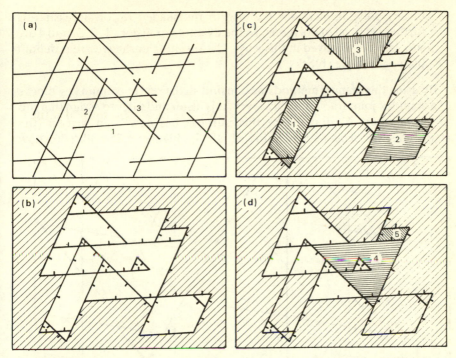

Figure 6.24 Generation of maximum key block regions from trace maps: (a) a given trace map for a free surface; (b) and (c) intermediate stage of deleting non-key block regions; (d) the maximum key block regions of the trace map, for the given JP, indicated by the unruled area.

a **map code** in which 0 and 1 mean 'above' and 'below' the joint trace respectively. Then, for each critical JP, there will be corresponding **map codes** indicating which side of the joint trace contains the potential key block. For example, in Figure 6.24, with three sets of joints, the map code selected for delineation is 011, meaning that the blocks of the selected JP lie above any trace of joint 1 and below any traces of joints 2 and 3. These sides of every joint trace segment have been labelled with a tick mark in Figure 6.24b. A real potential key block is a polygon in the map such that each boundary edge has a tick mark pointing into the centre of the polygon. The real maximum potential key block is the largest such polygon in the map, which will generally be formed by the union of individual polygons.

(3) *Delete all blocks intersecting the boundary*. Assuming that the edge of the map is the limit of a free face, and that the JP code we have selected belongs to the free face only and not to an edge, any polygon that intersects the edge of the map is not removable. In Figure 6.24b, the region outside the polygons has been shaded. Then shade any adjacent polygon having a tick mark pointed into the shaded area. The removal

of any polygon increases the size of the shaded region necessitating further removal of polygons. In the example, polygons 1, 2, 3, and 4 are successively removed in this way. The resulting unshaded areas delimit the potential key blocks.

By using this manual method one can find all the real maximum key blocks of a complex joint trace map; however, it is faster and safer to assign that job to a computer. A program for this purpose has now been written (the algorithms are explained by Shi *et al.* 1985). Figures 6.25 and 6.26 show

Figure 6.25   A trace map generated by joint statistics.

example computations. Figure 6.25 is a joint trace map, with tick marks on each joint corresponding to an input map code. The joint trace map can be input dirctly with the coordinates of the end points of every trace. Alternatively, it can be produced by joint trace simulation, using any one of three models. The most complete simulation model employs a three-dimensional array of joint discs of determined orientations and randomly assigned spacing and extent (Chan & Goodman 1983); a joint trace map is realized by intersecting the model volume with a specified excavation surface and the

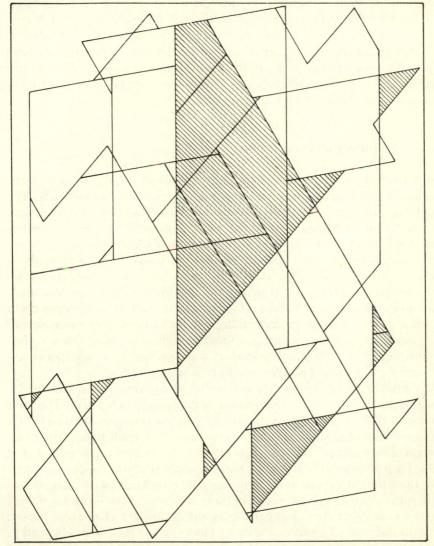

Figure 6.26 Automatically determined maximum key block regions for the trace map of Figure 6.25.

joint traces are the chords of intersected joint discs. A second model is simply a two-dimensional version of the first, with the three-dimensional joint information projected into the map plane and the simulation of linear traces performed directly in the map plane. The third method is a speedy, one-dimensional procedure in which a regular pattern of equally spaced, parallel line segments of equal length and equal separation (gap width) is perturbed randomly to have a distribution of lengths, spacings, and gaps and a fixed orientation. Figure 6.25 was produced in this manner. Application of the microcomputer program to Figure 6.25 yielded the maximum key block regions shown by the ruled polygons on Figure 6.26. It can be proved that the blank polygons determined manually in Figure 6.24d, or the ruled polygons located by the computer program in Figure 6.26, are the surfaces of a real, three-dimensional key block if the boundary joints of the polygon are sufficiently long. The proposition is obvious for a single convex block but requires proof for a complex, united block.

6.22   Summary and conclusions

This chapter introduces ideas for the analysis of infinite block systems created by the regular repetitions of joints of several different sets. The methods are rigorously derived from the principles of three-dimensional geometry and topology. Using them in an engineering project establishes the dominance of geological structure, rather than the stress–strain response of the material, for the stability analysis of excavations. The methods discussed use kinds of input data that can be fairly well established from preconstruction geological investigations, the primary input being the orientations of the joint sets. These methods are, above all, three dimensional in nature, designed to deal fundamentally with the three-dimensional nature of the blocks formed by the intersections of joints and excavation surfaces. Nevertheless, a number of principles fundamental to block theory are illustrated by means of simple two-dimensional examples.

A number of arguments were martialled here to demonstrate the fundamental and overwhelming importance of the joint pyramid (JP). The joint pyramid is the volume generated by intersecting one half-space each of a series of joint planes, each of which passes through a single point, termed **the origin of coordinates**. Every system of joints determines a given number of JPs. Each JP belongs to an infinite number of physical rock blocks, considering the different ways in which a volume can be enclosed by combinations of the particular half-spaces. Yet all of these rock blocks can be said to belong to the same block type; in other words, the JP concept effects a geometric classification of all kinds and sizes of blocks in the rock mass. Instead of operating on all the myriad individuals and unions of blocks, it is sufficient to operate on the joint pyramids alone in order to determine the removability and finiteness of blocks and block unions. Furthermore, the joint pyramid has the physical interpretation of being the collection of all sliding directions associated with blocks formed from that JP. Thus there is one and only one stability analysis for a given JP and this analysis can be used to discuss the support requirements for all the innumerable blocks associated with the JP.

Previous publications have dealt primarily with single convex blocks formed from the different JPs. Here we address the formation and analysis of united blocks. A united block is the union of any number of convex sub-blocks. It is generally non-convex. The intersection of an underground opening with a system of joint sets usually produces a number of complex, united blocks on the periphery of the excavation. The finiteness, removability, and stability analysis of united blocks are all controlled by the JP, as they are for convex blocks. In this chapter we have introduced the concept of a minimum stable united block corresponding to a JP that yields stable blocks under self-weight alone. If other forces combine with self-weight, because of the effects of water, inertia, or structural load, small blocks may become unstable whereas very large blocks are essentially unaffected. Thus there is a minimum stable block and all smaller blocks are unstable. Support of the largest unstable block in such a way that all its component blocks are also stable assures that the entire infinite system is safe. Support can be achieved using either of two principles: (1) force action of active supports, like bolts or cables, which turn the resultant into a favourable direction; or (2) constraint action of passive supports, like concrete piers, that prevent slip on the joints they traverse.

Clamped united blocks are those physical blocks that contain a set of parallel faces formed by joints of the same set. Unless such joints are exceptionally smooth, or filled with clay, or the rock mass is under a state of very low compression or actual tension in the direction of their normal, clamped united blocks are safe.

The removability of a united block is expressed formally by the necessary and sufficient condition for finiteness that EP and JP have an empty intersection. We have shown how to determine EP and JP for a general united block having complex systems of both bounding joints and free planes. The JP for a united block is the **intersection** of JPs for each of its component sub-blocks. The EP (excavation pyramid) for the united block is the **union** of the EPs for each group of free surfaces bounding a component block. The component blocks are created by extending either some of the joint planes or some of the free planes or both, as shown in a number of the illustrations. Using this theorem, it is established that a united block cannot be removable in an excavation when the block is bounded by opposed free surfaces of the excavation perimeter. It is also established, using the united block decomposition, that a single set of joint planes cannot create removable blocks. However, a joint system formed by a single set of long, smooth joints can create hazards for underground excavations driven in the direction of their strike if there is low normal compression in the direction normal to the joints.

In recent years there has been considerable interest in the use of classification systems for rock masses to assess tunnelling designs, before or after excavation as the case may be. Without entering into the discussion of the suitability and wisdom of embracing this approach for practical rock mechanics, it is possible to make one observation based upon the material presented in this chapter. Block theory shows that one of the most important factors determining the support needs of jointed rock in excavations is the number of joint sets. When the excavation is formed by only one free surface, at least three joint sets are required to create removable blocks. Compound free surfaces can reduce this number. But it is apparent that the number of joint sets in a rock mass is an important property of the rock with respect to excavation stability. This factor is a component of the Q system introduced by Barton *et al*. (1974).

The concept of a maximum key block region was established and examined for surface cuts, tunnels, and underground chambers. The maximum key block region provides a conservative upper bound for support requirements of united key blocks. For equivalent service widths, tunnels have much smaller maximum key block regions than do surface excavations, and therefore in rock masses with many joint sets correctly oriented tunnels are safer and more economical than open cuts.

If a joint trace map is made during excavation of an underground gallery, it is possible to establish real maximum key block regions that are considerably smaller than the maximum key block regions referred to above. Then considerable economies in support cost are realizable. A probabilistic analysis of support needs can be made by developing a model of the joint system to permit Monte Carlo simulation of trace maps; block theory can be applied to the simulations to achieve a series of real maximum united key block regions and corresponding distributions of support requirements. Considerable progress has already been made on computer generation of trace maps and automatic delineation of their maximum key block regions.

Using the concept of maximum key block regions, it has been possible to present the complete analysis of all JPs on a single diagram. This diagram and this type of analysis can be used to select an optimum tunnel orientation and an optimum support design. It can also be used to select an optimum separation between parallel tunnels or shafts, based on the requirement that movement of key blocks must not be permitted to undermine the whole width of the rock pillar between openings.

References

Barton, N., R. Lien and J. Lunde 1974. Engineering classification of rock masses for the design of tunnel support. *Rock Mech*. **6**, 189–239.

Chan, L. Y. and R. E. Goodman 1983. Prediction of support requirements for hardrock excavations using key block theory and joint statistics. In *Rock mechanics – theory – experiment – practice*, Proc. 24th US Symp. Rock Mech., College Station, Texas, 557–76. College Station, Tex.: Association of Engineering Geologists.

Goodman, R. E. and Gen hua Shi 1985. *Block theory and its application to rock engineering*. Englewood Cliffs, NJ: Prentice-Hall.

Shi, Gen hua, R. E. Goodman and J. Tinucci 1985. Application of block theory to simulated joint trace maps. In *Fundamentals of rock joints*, O. Stephansson (ed.), 367–83. Luleå: Centak Publishers.

Index